InfoWar in Cyberspace: Researcher on the Net

Published 2001
2nd edition

Published by Bruce Gold
360 -188 Douglas St, Victoria, BC.
Canada, V8V 2P1
goldb@Shaw.ca

InfoWar in Cyberspace: Researcher on the Net

Contents

Introduction 4
The Internet 11
  Origins of the Internet 11

  Dr. Strangelove 11

  Natural Democracy 17

  Commercialization 21

  Information Superhighway 25

  The Internet as a Technological Utopia 27
The Internet: Its Shape and Uses 31
  Shape 31

  Uses 33

  Internet's Impact 34
Knowledge and InfoWar 40
  Some Basics 41

  Knowledge as a Social Structure 45
Computerization and Thinking 58
  Packaging Knowledge – Packaging Thinking 58

  Software and Knowledge 64

  The New Economy and "Thinking Machines" 67

  Information and Ideas 71

  Ideas: Different Approaches 74

Fantasy of the "Thinking" Computer 76

Quantity Effects Quality 79
    A Tidal Wave of Data 79

    Information Mobilization... 82

    Impacts of the New Technology 84

    Problems of DataGlut... 88

    Nicheing 90

    Cultural Impacts of the Information Economy 93

Censorship 97
    General Nature and Purpose.................................. 97

    Shaping Society.. 104

    Censoring the Net.. 106

    Can the Net be Censored? 110

Research 114
    Theory... 115

    Ontology 117

    Epistemology... 118

    Assumptions.. 119

    Causation.. 121

    Evidence... 122

    Measurement Systems.. 126

    Logic 128

    Statistics.. 129

    Dis-information 131

Bibliography 139

Introduction

This book is about the Internet and the struggle to control it. It explores the political and intellectual issues that play out in Cyberspace. The book also addresses the ideas and concepts that underlie and inform the research process. It offers the researcher intellectual tools and an understanding of the research process. It gives the researcher the background necessary for doing research in the contentious and contested waters of the Internet.

So, let us begin with the basic issue. The new, and much hyped, "Information Age" has radically democratized communications. Never, has so much communications ability been available to so many. Those of you familiar with the hype will probably note that I did not say "radically democratize information" or "knowledge" or "access to information." Despite millions of words and millions in advertising dollars the Internet is not about an "information revolution." It's about a revolution in mass communications. To understand this rather contrary statement let's look at some "revolutions."

The Revolution in information storage and transmission – writing - happened several thousand years ago. It overcame the limited storage ability of memory and the dreadfully error prone and chancy distribution of oral information. It also overcame the survivability problem of oral communications.

The Cost Revolution in information storage – printing press and the book – happened about 200 years ago with the development of cheap printed material. This brought information to a broad, even worldwide audience. Cheap reproduction of information opened the era of political broadsheets (18th century) and continued into the era of the "yellow press" and "muckrakers" (19th- 20th centuries). The public library movement (started in the 1890's) brought this information to the masses.

The Revolution in communications speed – telegraph, telephone – happened about 100 years ago. The speed of information moved to same day – anywhere in the world (if it was on the network!). However, it was limited to individual to individual messages.

The Revolution in knowledge (the knowledge explosion) – The deliberate invention and funding of information factories as a specialized endeavour; research labs, research universities – happened in the early 20th century and massively accelerated during WWII.

The Revolution in broadcast communication's speed – radio, TV – happened about 80 years ago for radio (as a mass medium) and about 50 years ago for TV. Anyone with the right equipment could receive broadcast information. However, cost of equipment and the limited number of channels (bandwidth) restricted choice and limited information to that chosen by the owners of the medium.

The Revolution in mass communications – Internet- began about ten years ago. It established cheap, worldwide communications between individuals and between individuals and a mass audience. For little more than pocket change anyone or any group can put up a web page, engage in a little selective advertising and reach an audience of hundreds of thousands.
The Internet is something new in the history of the world; a vast electronic commons where more people

than ever before have a voice. It's also, at least currently, an uncontrolled mass media. Radio, TV and mass publications are all controlled by their owners. Owners, who obtain influence by gate keeping what information is made available and how that information is presented. Historically, this control has been a tool of governmental and elite power and continues to be so. Information, control of discourse and control of perception are often the keys to power. The Internet, by being open to all, is a potential threat to and support for the powers of special interests. However, without elite control of content, people using the Internet must learn to sort out the information for themselves. This is the challenge that goes with an open marketplace for ideas and information – people must develop the ability to cut through the rhetoric and make reasoned judgments.

I'm a professional researcher. I have a Master's Degree in Intellectual History (the study of people and ideas and how they interact). I also have a Master's Degree in Public Administration where I studied policy and policy making. Over the last ten years, I have watched the Internet grow from a weak and trivial source of information to a massive sea of information. I have also watched it become an arena for agenda scholarship. What is agenda scholarship?

Agenda scholarship is the biggest problem faced by those doing research on the Net (and elsewhere). Agenda scholarship is the intellectual equivalent of reverse engineering. One starts with the policy outcome that one wants. Then you decide what "situation" would justify such a policy. Then one constructs a "research study" to "find" the situation one desires. This is then re-packaged as "objective" research that justifies the desired policy. Politicking for policy change then follows based on one's "scientific facts and figures."

This book is about research and understanding how research works. It exists because I'm tired of seeing well-intentioned people, particularly inexperienced researchers, misled by clever manipulators masquerading as scholars and social scientists. Let's look at an example of agenda scholarship.

Some people believe that civilian gun ownership and use should be restricted; especially arms that might be used for self-defense. In order to justify their preference for a dramatic decrease in the right to arms, this group needed to "prove" or "demonstrate" that these arms were not needed. They needed to show that civilians who attempted to keep or use arms for self-defense were deluded about the usefulness of guns for self-defense. They wanted to prove that guns were so specialized, problematic and downright dangerous that the average person was incapable of using them to deter crime. Indeed, that the average homeowner was a menace to family and self. To do this they needed to design a research study that would demonstrate a very high level of failure in self-defense and a very high level of accidental shootings. The study could then be used politically, to justify the cancellation and/or downsizing of the right to arms and impose extensive restrictions on their use - especially for the purpose of self defense or crime prevention.

The study was completed and presented as scientific research, with all the normal professional methodology and statistics. It demonstrated that guns are amazingly useless for self-defense. It found that a homeowner was 43 times more likely to kill themselves or an acquaintance than to kill an attacker (Protection or Peril: An analysis of firearm-related deaths in the home by AL Kellermann, and DT Reay). Supporters then used the study to justify stiffer gun laws and to categorize guns as instruments suitable only for sport or crime. However, the study contained carefully designed flaws and incorrect assumptions that totally skew the results towards the desired outcome. It is, in short, a classic example of agenda scholarship.

The study used the survey method to find households were a homeowner had used a gun for crime prevention or an accidental shooting had taken place. The author them examined these instances to see if gun ownership made homeowners safer or more at risk. So far so good – but the author wanted to prove how useless and dangerous guns were so he stacked the deck. To begin with, he chose "shot dead" as his criteria for a successful defense. Second 83% of the sample was suicides. Lastly, he used the FBI definition of acquaintance.

How does this skew the results? It does so because it uses an incorrect measure. Homeowners seldom kill an intruder. In fact, official figures demonstrate that in 98% of defenses civilians do not fire the gun at all. The mere presence of a gun deters the criminal. By choosing shot dead as his criteria for success, the author immediately eliminated over 98% of all successful gun defenses! (We might also note that by doing this he also presented fatal shootings as the "norm.") The use of suicides is also an error. This is not a study of suicide; it's a study of householder use of guns for crime prevention. The author should have deleted these suicides from the study. (In fact, there were so few legitimate examples of a homeowner or acquaintance killed that the study's results were statistically insignificant- too few to be distinguishable from background "noise.") The last maneuver, the use of the FBI definition of acquaintance is an exercise in misrepresentation. The nonprofessional assumes that acquaintance means friend, relative or neighbor. The FBI definition merely means someone known to the person. Thus, two drug dealers who knew each other shooting it out would be an example of an "acquaintance" being killed.

We could at this point simply say, "bad social science" or "methodologically flawed" and junk the whole thing. However, to understand what is happening we need to know more about knowledge, thinking and research. (This book is not about the various positions one could take in the debates mentioned, it is about understanding what writers are doing and why.)

The Internet is badly infested with this sort of agenda scholarship. They look like science, but are in fact polemics (political arguments designed to persuade). This book is an attempt to give inexperienced researchers the intellectual tools they need to sort agenda scholarship from true scholarship (we're talking quality, not if it comes from a University).

The plan for the book is simple. First, it rejects the "cook book" approach of simple "follow the steps" procedures. It also rejects the "list approach" where research is conducted by referring to a list of rules, or lists of methodological errors. These approaches can be useful, but one soon finds that the steps don't fit the problem or that the lists have become too long and complex. What a researcher really needs is an understanding of principles and structure. How does knowledge work? How does theory and research work? How do propaganda and the mass media work? This book addresses these issues in the belief that if you understand how things work you can sort out what is going on.

The book begins with a description of the Internet, its history, its impact and its issues. This is important knowledge for a researcher using the Internet and sets the context for the rest of the book. The second chapter examines how social groups create and use ideas. This gives the reader a basis for understanding the characteristics of InfoWar. The third chapter examines computerization, which is the technological foundation of the Internet. More importantly, it examines how computer enthusiasts have distorted our approach to thinking and analysis. The fourth chapter looks at the problem of information overload and its effect on research. The fifth chapter examines the issue of censorship and censorship's impact on research. The last chapter builds on the first five chapters and the intellectual base that they create. It examines how research is put together and how it "works." Throughout the book numerous comments are made on how the material is currently being used and misused. The book concludes with an examination of dis-information, a growing problem for the modern researcher.

In doing this, the book makes every effort to keep the material clear, stick to the basic issues and avoid confusing side issues and complications. It would be pleasant to claim that the result was an "idiot's guide" that anyone could master based on a few simple rules. Unfortunately, this is not possible in an area that is both complex and contested. However, the book establishes a basic intellectual map, it outlines the issues and it points to particular difficulties and dangers. I have tried to spare the reader from academic language and academic nit picking. For those willing to undertake the journey the book will "put you on to the game" and give you the tools to dig deeper and understand better.

The Internet

Origins of the Internet

To understand information problems encountered on the Internet and the issues raised by the Internet. It's necessary to understand how it came to its present structure and organization. The first, and essential, understanding is that the Internet is constructed on contested ground. Its history is a history of conflicting political and social agendas. This history, complex and filled with both intentional and unintentional events, has been described by three popular stories. Understanding the Internet (and its uses) requires the understanding of these three overlapping versions of its genesis.

The first story is the story of Dr. Strangelove's computer and a decades long military effort to win the Cold War. The second story is the story of the "natural birth" of a technology and its inherent love for democracy and individual freedom. The third story is one of "free enterprise" and the forces of private capitalism establishing a new technology, indeed ushering in the new economic order of the information age. All three stories have elements of truth in them and all three stories are slanted to define a particular vision of what the Internet is and thereby form a vision of what it should be used for. This chapter examines the history of the Internet and the stories tied to that history.

Dr. Strangelove

The Internet began as an instrument of war. It was an instrument designed and intended as a weapon of victory over a relentless foe. The strategy underlying this event had much to do with top down command and control and little to do with democracy. As a strategy, it had a complete and absolute contempt for economics, much less market economics. The Internet's origins lie in the computer projects of the Second World War. At that time, facing the complex problems of warfare technology and the massive logistics problems of a World War, the Allies invested heavily in the infant technology of the computer. This investment, wholly uneconomic, led to the modern computer, the transistor, and a host of other technological marvels. This investment continued into the post war era. In the 1950's, the military arm of the US Federal Government provided 80% of all computer development funds. Government was the only market for computers and it was a predominantly military market. This government funding, a classic case of infant industry protection and development established American hegemony in computer and related technologies.

In 1957, the national panic over Sputnik stimulated these military's projects and led to the founding of the Advanced Research Projects Agency (ARPA) of the Pentagon. Amongst other projects, the agency explored the idea of expanding its computer ability by linking computers into a network. This would make the best use of the expensive computing power of the era. It also inspired a vision of vastly increased command and control capabilities based on a network of linked computers. However, to achieve these very desirable outcomes, network designers would have to overcome a number of critical technical problems. Simply put (and vastly simplified), the three most pressing problems were the problems of transmitting error free messages, the difficulties in linking computers – and sub networks – using different codes and the problem of survivability.

The first problem, the problem of message reliability was critical because any system of linked computers was only as good as its ability to pass error free information. This difficulty increased with the length of the message and the number of messages sent. This issue posed an increasing threat to systems reliability as the network increased in size. Researchers found that this problem could be addressed by breaking messages into smaller parts called "packets" and then sending each packet as a separate message (technically this is called packet-switching). The first successful demonstration of packet switching between two computers was carried out at the National Physical Laboratory in Great Britain in 1968 and opened the possibility of a practical computer network. As we shall see, this process of packet switching would have an unforeseen structural impact on how the Internet would eventually be organized.

Within a year (1969), the Pentagon implemented a computer network using the packet-switching technique. The first node in the system was at UCLA and the system was expanded to four nodes by the end of the year. This creation was named the Advanced Research Projects Agency Network (ARAPANET), the ancestor of the current Internet. This initial linking was the first step in a process that promised a more efficient use of the very expensive computing capacity. However, the Pentagon was interested in more than a network of research computers; it also wanted a bombproof command and control capability.

This need, quite literally the threat of nuclear annihilation, put the Pentagon in a very painful strategic dilemma. The post-WWII military leadership was keenly aware of the importance of security and the importance of maintaining a rigid chain of command. Control of information and control of who was allowed to talk to who (lines of reporting) were classic, time-tested methods of enforcing security, obedience and control. To violate these principles, especially in the face of the era's massive anti-war protests and civil disobedience was clearly not an attractive option for the military. However, these officers were also deeply familiar with the immense disruption caused by strategic bombing during WWII. Faced with the brutal possibility of strategic nuclear bombing orders of magnitude greater than WWII the Pentagon was forced to give precedence to survivability over all other considerations.

The Pentagon weighted these dilemmas and made a radical break with its traditional command and control strategies. Despite misgivings about security and about keeping information locked within the command structure the Pentagon de-centralize the net. Following a proposal of the Rand Corporation (first made public in 1964), the Pentagon designed the network without a central authority. Instead of the usual pyramid of nodes with rigid report lines from lesser too higher headquarters, the nodes would have no hierarchy. Each node (computer) would have the authority to originate, pass on and receive messages. Each message would be divided into packets (each with addresses and coding to detect errors) and would make its own individual way through the network and be re-assembled at its destination. By eliminating the pyramid structure of a formal chain of command, the Pentagon also eliminated information bottlenecks at each headquarters. This also eliminated the normal report lines and allowed anyone to talk with anyone.

This was radical stuff for a traditionally conservative and hierarchical military. However, when they moved from an information pyramid to an organization more resembling an information fishnet they were able to make the system bombproof. It was designed so that damage would simply be bypassed and no enemy could behead the system with a single hit on the top of the pyramid. However, as we shall see, by eliminating the information hierarchy they allowed everyone to talk to everyone (information democracy) and prevented anyone from filtering anyone else's information. All information on the Net became available to all members of the Net.

ARAPANET had achieved both signal reliability (packet switching) and bomb proofing (decentralization), but still suffered from other communications problems. It existed but at this point in its evolution (early 1970's), it was still a very small and closed system tightly controlled by the Pentagon. However, this was changed by the continuing strategic imperatives of the Cold War. Cold War politics of the 1960's, stimulated by the Cuban Missile Crisis, stressed the need for a national command and control network. (This vision was later to enter the national consciousness through such movies as Dr Strangelove). This drive for the widest possible command and control network required a dramatic expansion of the small University based ARAPANET. The military also wanted to include its new packet-switching technique to radio and satellite networks. However, linking these government networks and incorporating the growing number of civilian computer networks into one system presented serious difficulties.

The basic problem was the independent development of multiple networks with multiple hardware configurations and multiple software systems. This created technological incompatibilities that resisted networking. Indeed many private, commercial networks had pursued a deliberate strategy of incompatibility to lock in customers and lock out competitors. (This strategy is known as "technological capture," it makes customers dependent on your system's privately owned technical standards to prevent them from using the services of competitors.) One approach to solving this problem would have been the imposition of a universal computer language. Unfortunately, this solution was neither technologically feasible nor politically possible. The alternative solution was the development of an electronic lingua franca or shared language that everyone could use for communicating on the Net.

Despite continuing concerns over security and control, the Pentagon followed up its decision to decentralize with the development of a common Internet language. This language was left in the public sector for anyone to use. This common communication standard is universal to the Net and handles the addressing and routing of packets across multiple computers and networks. This first common electronic language the Network Control Protocol (NCP) was since been superseded by the current Transmission Control Protocol (TCP) which converts messages into packets and then re-assembles them at their destination. Another common code the Internet Protocol (IP) handles the addressing of messages. This common language, TCP/IP is the lingua franca of the Internet. Technologically, this development of a publicly owned communications standard increased the democratic character of the network. Nobody could stop participation by denying use of the common language or restrict entry by price gouging for code usage.

By a curious twist, the desire for a top down command and control system had produced a "democratic" communications system. The Cold War shaped the development of computer and Internet technologies; but these developing technologies also shaped the Cold War. Because of the military's desire for a centralized control system, no effort or expense was spared in research and development. This desire also created the technological fantasy of a "super brain" that would give leaders real-time control on a gigantic scale. This vision supported the dream/nightmare of world politics as a system of technological management. Cold War politics were being institutionalized within its machines, which in turn made possible the centralized geopolitics of the Cold War. As a result technology and a particular group's vision of technology had shaped both the technological and the political world.

The military formally withdrew from the Internet in 1983 when they separated their systems off into an independent MILNET. However, military concerns remain and surface as encryption policies and in programs such as ECHELON, the worldwide spy network that monitors all broadcast communications.

Natural Democracy

The genesis of the Internet's is also the bottom up story of how the rebellious 1960's whiz kids at the universities hijacked the ARAPANET for their own use and over time shaped it to serve democratic purposes. The first problem for this democratic agenda was the limited distribution of ARAPANET sites. The system, founded in 1969, had only connected thirty-seven universities by 1972. This was a very restrictive distribution in a University system with rapidly expanding computer science departments.

The system also suffered from lack of use. In 1972, the International Conference on Computer Communications in Washington had publicly demonstrated the success of packet switching and the technical success of ARAPANET. Unfortunately, the idea was so new that use of the Net was only running at about 2% capacity. In 1972 engineer Ray Tominson developed software for sending e-mail and by 1973 three quarters of network traffic was e-mail. The system, designed to link computers for research, was being used as an electronic post-office to carry news and personal messages. The users had spontaneously invented e-mail and happily subverted the system to their personal use. Despite the top-down intent of the network, the actual users were shaping it into an on-line community.

Through the 70's, the ARPANET grew as other networks and machines came on-line. Its decentralized structure and common language allowed the inclusion of many different kinds of machine irrespective of contents or ownership. This led to the development of the mail-list, a broadcast technique allowing multiple recipients of a message. Again, despite opposition from system's administrators, the messages branched away from official subjects and official channels (one of the first big mailing lists was for science fiction lovers).

This development of mailing list discussion groups led two Duke graduate students Tom Truscott and Jim Ellis to develop some simple software for UNIX (a very common operating system developed in the 1970's by Bell Laboratories). This software allowed computers to link up and exchange files. This 1979 software innovation permitted on-line newsletters that could be read and added to by every computer on the system. The same system allowed e-mail to be sent between UNIX computers connected by telephone modems. This development was seen as a poor man's ARAPANET, available to anyone with the requisite computer. The upstart network, named USENET, grew from three initial sites to 150 two years latter, to 11,000 sites in 1988. (Currently USENET is part of the Internet

and comprises some 30,000 different news groups.) A
further split occurred when the administrators of the
USENET refused to carry newsgroups for sex and drugs.
Usenet participants responded by setting up the "alt."
distribution network on the decentralized net. Alt.sex
and Alt.drugs were the first alternative newsgroups.
(Closely followed by alt. Rock-n-Roll!) The USENET
system linked into ARAPANET in 1981 when a graduate
student, Mark Horton, established a one way gateway. A
fuller linkage, essentially merged the systems was
established in 1983 after the military lowered security
concerns by splitting their own MILNET off from
ARAPANET.

Thus a computer network created for military/political
command and control became a very democratic means
of exchanging messages and a mass communications
media. A development directly related to the 60's
demand for more democracy. The public nature of the
TCP/IP code and the decentralized nature of the
network allowed anyone with the requisite machinery to
hook in, creating a true electronic commons. Since
government had already covered the investment costs
and each node (user) paid their own expenses additions
to the Net cost the Net nothing. At the same time, they
made the Net a more valuable means of communications
and a richer source of information. This "democratic"
network of "uncensored" debate has been seen as the
successor to the people's presses and broadsheets of the
American Revolution and the Penny Presses in England.
However, this development was only open to those who
owned computers, a very select group in the 1970's and
80's. (This concern over access is present in
contemporary debates over the existence of a "digital
divide" between those wealthy enough to own computers
and those who are not.)

In 1984 the National Science Foundation's Office of Advanced Scientific Computing formed the National Science Foundation Net (NSFNET) and began a program of rapid research and development. With other government agencies, such as NASA, major advances in computer, software and networking technology led to rapid increases in network capabilities. ARAPANET formally closed in 1989, was succeeded by the current Internet. By 1993, the government had linked the Internet with America's future and the National Information Infrastructure (NII) initiative was established to increase Net capacity and speeds by orders of magnitude. The new medium became identified with progress, competitiveness and all future success. Official pronouncements predicted that the new, improved Network would "unleash an information revolution that would change forever the way people live, work and interact with each other." The Internet, as a vast electronic commons of communication and a new decentralized system of mass communications was officially established.

The remaining part of the "democratic technology" story is the story of the personal computer that supported the average person's access to the Internet. Here again it was radical technologists, who created these inexpensive alternatives to mainframes. In 1970, Felstein of the "Homebrew Computer Club" conspired with other computer hobbyists and computer liberationists to create the first PC's. The arrival of the relatively inexpensive PC and the relatively user-friendly software to operate it established the last technological link for the establishment of a very broadly based electronic communications commons.

Commercialization

The third story of the Internet is the story of commercialization. As early as 1959, Daniel Bell, a Harvard sociologist, had identified information as a commodity. This was an idea that easily evolved into the idea of the Internet as a market. In addition, like the Internet, the idea started with computers. The first personal computer, the Altair, appeared in 1975. Then when two amateur (and teenage) programmers, Bill Gates and Paul Allen, wrote a software program for it, the informal hobbyist/hacker culture assumed their usual attitude of "information wants to be free" and immediately began to distribute free copies of the program. This led to an angry response from Gates that even thought the software language he was using (BASIC) was free his software creation was property, which entitled him to payment.

In 1972, ARPA tried to sell the Net. The major telecommunications corporations, including AT&T, showed little interest. However, large corporations such as BBN (TELENET), IBM (SNA) and Digital Equipment (DECNET) began building their own proprietary networks to lock customers in to their systems. These proprietary projects were defeated by the government's commitment to "open standards." However, with little political will to maintain the Internet as a public infrastructure and with increasing commercial interest there was a steady shift from public to private ownership. The anti-establishment tendencies of the 1960's were morphing into anti-government tendencies of the 1980's and 90's. In practice, this shift transferred the benefits of huge public investments into private hands.

In the 1980's the National Science Foundation (ARPA's successor) moved to privatize the Net. This privatization was part of an immense infrastructure development sparked by the combination of cheap computing, cheap information storage and cheap communications. By the 1990's, computing and telecommunications investments were fully half of all American capital investments. Worldwide, the 1996 corporate/ government spending on Information Technology reached $1,076 billion, with US expenditures around $500 billion (2.8% of US GDP). This shift towards private ownership was accompanied by a shift towards private control of information. Following a pattern set in newspapers, radio and TV there was a shift towards corporate domination of the Internet.

This move towards a mass media of controlled and filtered information is particularly evident in the proliferation of on-line think tanks. These think tanks originally came onto the political scene as a conservative response to the "to left - too socialist" ideas that dominated the Universities in the years following WWII. In the 1980's right-wing think tanks like the Hoover Institute, Heritage Foundation, American Enterprise Institute and the Center for Strategic and International Studies proliferated. Many of them got their funding from right-wing corporate foundations bankrolled by big-money families. These organizations consciously and explicitly attempted to control the country's political agendas by supplying information, shaping debate and presenting arguments. Left wing think tanks, indeed political sites of all persuasions, have also benefited from the Internet's low costs; but are a much smaller and less influential portion of the Net.

As large commercial enterprises expanded into the Net, they were primarily concerned with profits and sustaining the laws and policies that support their profits. These focuses lead them towards the same strategy of dumbing down of content to scandal, celebrities and sensationalized events and reducing information infotainment. They also attempted to maintain their role as information gatekeepers. On occasion, this would lead them to criticize government. However, when government policies meet corporate approval support was forthcoming. Some recent examples of this were the media's willingness to promote a perception of Mexico as a "true democracy" just before the NAFTA vote and their acceptance of the "stolen incubators" propaganda stunt before the Gulf War. Not surprisingly, a recent report by the Pew Research Center has found that 20% of Americans are using the Internet as a news source at least once a week – three times what it was only two years previously.

This trend towards commercialization and corporatism is very much disputed ground. Independent and anti-corporate sites are proliferating on the Net. Scholars, both official and otherwise, are expanding their on-line activities and reaching larger audiences than ever before. At the same time that these independents are proliferating major players such as AOL (America On-Line) are subtly limiting the Net. This limiting comes in part from a desire to market the Internet by simplifying it for new and unsophisticated users. It also comes as the direct result of commercial sales. "Simplifying" the Net dovetails with the desire to shift its contents towards commercial venues and deliver information aimed at consumers. The creation of cheap "network computers" that allow access without much in the way of information handling abilities furthers this trend. It has been described as reducing the Internet to the level of a consumer oriented "interactive" TV.

This move extends beyond mere "service." Major on-line servers such as AT&T and AOL are now providing "edited" directories, theme areas and links. These pre-selected sites are presented as "preferred sites" chosen for some particular merit. These sites direct the user away from any site critical of corporate interests or sites that corporate interests deem "too controversial." It's also common for such listed sites to pay for the privilege of being listed. It's unclear how many new or unsophisticated users realize that their Internet access has been filtered, directed and purchased. Advertising revenue streams are part of the web site design. Many users do not realize that unlike an Internet connection and a web browser these highly selective corporate "gateway sites" to the Internet are unnecessary.

This race for commercial dominance has also led to rapid improvements in the rate of data delivery (often referred to as increasing bandwidth). In general, this trend is beneficial since it speeds up access and allows the transmission of movies and other demanding formats. However, this trend also shifts the Net in the direction of increasing and increasingly expensive production values. It has become commonplace to see ads touting fancy graphics with the implication that an increasingly flashy form will somehow increase the value of content.

These elaborate corporate web sites enjoy huge marketing and advertising campaigns. For many users, especially the new and unsophisticated, a corporate owned "gateway" site shapes and directs their access. Despite this trend, the decentralization of the Net continues to shape its usage. On the Internet, all addresses are equal (unless one is accessing through a "preferred link!"). The non-commercial addressing system of the Internet gives equal billing to a personal web site set up for as little as $30 a month or a corporate site costing hundreds of thousands. Consequently, the Internet can be said to be going in two directions at once. It's moving towards a more commercialized, more corporate controlled medium and towards a more open more inclusive "democratic" medium. In part, this trend is shaped by the knowledge and experience of the user. Some users are aware of the situation and conduct their own searches and assessments; some users are blissfully unaware that they are being channeled and directed. The Net experience is also deeply defined by self-selection. Those who are interested in content soon learn to ignore flashy graphics and glitz. Those who want infotainment find it.

Information Superhighway

In the early 90's, the Internet began to be presented as an "Information Superhighway," a term that evoked images of the great public works projects that became the interstate highway system. It suggested a public infrastructure available to all and for the benefit of all. The prevailing image was one of exploration, learning and shared communications resources. However, after vast amounts of public funds had been spent building the Highway the image began to change from an education/information infrastructure to an on-line marketing infrastructure. Part of this shift was promoted by the mass media's own presentations of the Net. According to the Nexis database, a database of news stories, in 1995 major newspapers referred to the "Information Superhighway" 4,562 times and electronic commerce 915 times. By 1999, despite a massive increase in non-commercial usage the "Information Superhighway" was only mentioned 842 times while e-commerce was mentioned 20,641 times.

Large corporations now own most of the large volume sites on the Internet. Search engines, the web browser software that locate information on the Net have becoming increasingly skewed by behind the scenes commercial arrangements. The new hero's of the Internet are not technicians or visionary pioneers; they are executives such as Bill Gates, Jeff Bezos and Steve Case. As the Net has been commercialized, it has also become more the creature of the same large corporations that dominate the print and broadcast media.

This commercial scramble, like previous developments of the Internet, has been stimulated by politics. The Clinton administration determined that private industry (with government assistance) would build and run the National Communications Infrastructure (NCI). As owners of the "gateways" and the physical infrastructure, commercial interests would shape this infrastructure. Under this, increasingly commercial, control the Net's increased bandwidth (the 5000 channel universe) might have little to offer other than the repackaging of re-run material from the 5 channel universe. Corporate profit and corporate ideas about appropriate market development will and are taking precedence in an increasingly commercialized Net.

Yet the Internet is and remains contested ground. It's already one of the world's greatest information sources. It is a mass communications medium currently available to the powerful and the relatively powerless, a first in history. Some powerful interests seek to extend this "democratic" medium even further. Some interests want it restrained for political and ideological reasons. Some wish it turned over to the media establishment in the same way that radio and TV (both public mediums) were turned over to the commercial sector. Who is going to run the show? Who will write the next "Internet story"?

Good questions. In the meantime, those using the Internet as a source of information need to be sophisticated about information and about the political forces shaping their information medium.

The Internet as a Technological Utopia

Technological Utopianism is the belief that technology in general or a specific technology in particular will be instrumental in achieving an ideal world. For example, the White House released documents in 1993 that envisioned the Internet as the basis for a world where people could live anywhere and telecommute to work. This vision also included universal access to on-line medical care and the best on-line education. How would this happen? Who would pay for it? What would the political implications be? These issues were left unanswered. Computer networking would do it all - presumably by simply making it technologically possible.

Technological Utopianism has been encouraged by Marshall McLuhan's 1960's predictions of a media driven future and by Alvin Toffler's Third Wave theory of a new Information Age. These visions were reinforced by ideas borrowed from science fiction writers like Heinlein and Asimov, who portrayed exciting futures filled with rugged techno-savvy heroes. These writers and the West Coast Intellectuals promoting many of their ideas saw the new technologies as a basis for Libertarian politics. Politics that would, in time, lead to a Jeffersonian Democracy. In this vision computers and the Internet would form the basis of an electronic "Commons" were individuals could express themselves in Cyberspace.

The convergence of media, computing and telecommunications would transform the social, economic and political world. Big government and big business would both be subdued and overthrown by the new technology's abilities to empower individuals. Leftist intellectuals like Howard Rheingold believed that the new technologies would allow individuals to exercise the extensive rights of speech and assembly (virtual assembly) advocated by America's founding fathers.

The New Right also embraced this technological vision arguing that the new technologies would stimulate private enterprise and create wealth. Combined with the removal of all regulatory restraints, this new Technical Utopia would empower the citizen (or the consumer) and result in a Jeffersonian vision of prosperity and political independence. Each member of the new "virtual class" would have the opportunity to become a successful entrepreneur. Faced with the competition of these new hi-tech pioneers the existing social, legal and political forms of the old (and over regulated) industrial era would go the way of the dinosaurs. The unregulated interactions of individuals in a completely free market would shape the new economy and the new economy would re-shape the political and social world. Government intervention was seen as an obstacle to this progress. The "invisible hand" of the free market and the almost Darwinian forces of technological evolution would usher in the new Information Age.

This vision of the future was expected to penetrate the factory and the office as well as the home. All industrial economies would need to wire their infrastructure to gain the competitive advantages of the digital age. Technological Utopianism saw the impact of unlimited quantities of information and previously unheard of communications capacity at essentially benign. The New Left and the New Right shared the vision of a brighter future. Both saw a decline of big government, big business and the intrusive nation state. For the Left, this new Jeffersonian future would be based on the freedoms of speech and assembly. For the Right, the new Jeffersonian future would be based on the freedoms of property and commerce.

This optimism, shared by both the political Left and Right, has tended to hide the very real problems associated with this future. The Right's vision of wealthy hi-tech entrepreneurs is balanced by the reality of low paid subcontractors and the insecure contract work that dominate the industry. The Left's Utopian vision of a new, clean, classless society ignores the widening gap between the have and have-nots. The immense increase in the wealth of the wealthiest has been balanced by decreasing wealth and harder times for the majority.

Both the economic liberalism of the Right and the social liberalism of the Left have glossed over the problems of a new Information Age. They have done this to the point of a willing acceptance of national de-industrialization. In this atmosphere of uncritical optimism, both Leftists and Conservatives have seen technology as a quick, easy road to fulfilling their dreams. The rhetoric of a Libertarian "Commons" and the rhetoric of a liberating "free market" struggle to define the character of Cyberspace.

In practice the "sacred tenets" of economic liberalism are completely at odds with the actual history of these "saving technologies." Private enterprise has played an important role but only within a framework of government funding and leadership. The Leftist vision gives little attention to the minimal gains and increased stresses that twenty years of technological revolution has brought to the majority of people. In practice, the new "Information Age" and "Knowledge Economy" are merely a reflection and continuation of the mixed economy that preceded them. Behind a shared rhetoric of anti-statism, both sides pursue the rhetoric of hi-tech libertarianism with (elements of hippie anarchism and economic liberalism) submerged in an overriding technological determinism. This technological determinism ignores the political and economic will that created the technologies in the first place.

Both capitalists and well-paid workers of the new hi-tech and information industries are unwilling to acknowledge the huge public investment that created their technologies, especially in the context of public under funding of health care, education and public infrastructures. An ever-expanding rhetoric of the "new marketplace" is increasingly used as a solution to the pressing social and economic problems of the day. Solutions free from the painful remedy of taxation or any repayment of the huge public investments that are the foundation of their prosperity.

Both Right and Left promote the new technology as the road to liberating individuals from the hierarchies of state and economics. In reality, the increasing polarization of American society is slowly undermining both their Utopias. The technologies of freedom are being shaped into technologies of dominance. The Internet, its control, its use and its ownership have become the battleground for liberating vs. dis-empowering technological developments, as well as the battleground for Left and Right ideologies.

The Internet: Its Shape and Uses

Shape

The spread of the Net in the 90's has resembled the spread of computers in the 70's and has given those computers a cheap means of interacting on a world scale. This worldwide Net is divided into different "domains" each denoting a different type of institution. Currently the Net is divided into .gov (government), .mil (military), .edu (education), .com (commercial), .org (non-profits), and .net (gateway computers serving as doorways between networks).

In part, this growth has been a reflection of society's increasing complexity and specialization. This growing social, economic and political complexity has been accompanied by an increasing need for information. Indeed, information has been described as the "oil" of the future, a strategic resource. This has led to the construction of the largest machine in human history, the global telecommunications network. A machine so large and complex, that no one knows its full size and shape or can track its daily even hourly evolution.

This global network is a composite of oral and written culture. Writing and the written word have become more accessible and more important. Yet the Net's immediacy, especially e-mail and "chat" forums, resembles the immediacy of oral communications. Lacking a formal hierarchical structure, the Net often resembles a tribal organization governed by not always clear structures of self-regulation.

This growing "machine" is promoted by those who have a commercial interest in its future. Their optimism has inspired many social critics and politicians to support the

Net as a "sure fire" solution to many of our problems. This prospect of a cheap, quick fix for persistent problems is very popular. For example, it's significantly cheaper and easier to connect a school to an Internet or put a computer in a classroom than to sort out a defective curriculum or incompetent teachers. This level of boundless enthusiasm has led to turf wars as various interests struggle for dominance and profit. Mergers, alliances and often fiscally unstable dot coms have proliferated as the commercial sector scrambles to get on the bandwagon of a "sure thing."

The increasing dominance of the traditional media (radio, TV, press) by fewer and fewer big players, (Five news agencies around the world control about 96 percent of the world's news distribution) is being balanced by an emerging technology that supports unprecedented opportunities for access and diversity. The Internet is rapidly breaking the old top down power of the traditional media's monopoly on information distribution. However, this development is contested and there are serious doubts whether the Internet will remain accessible, affordable and uncensored. Governments are beginning to impose harsh censorship laws targeting the Internet. (In the US, these laws have imposed morality (pornography) restrictions and information (drugs), restrictions that are not imposed on other media, particularly books.) Media conglomerates are increasingly gaining ownership of the fiber optic networks that support the Net and are already imposing their own "standards" on content. WorldCom's take over of MCI will give it control of more than half of the Internet traffic lines with the rest in the hands of corporate giants like AT&T, Sprint and GTE, Established government and corporate elites are, on the whole, pleased with the technological abilities of the Internet. However, they want these capabilities to be handed over to their control in the same manner that the once public capabilities of radio and TV where handed over to them. After half a century of enjoying the political power associated with media domination, they are not eager to see the Internet undercut their influence. The existence of the Internet, decentralized to the point of anarchy, open to anyone and capable of reaching a mass audience is a threat to their power. Like many who oppose their agendas they realize that technology is apolitical and will support the agendas of the people who control its use.

Uses

The Internet has come to be the servant of every cause. The commercial sector is rapidly expanding the interactive gaming, commercial infotainment and shopping at home possibilities of the Net. The government, in turn, has begun to assert how the Internet will create better schools, lower health care costs, create on-line libraries, and provide more efficient government services. Individuals are finding new opportunities to communicate and gain a public voice.

There are three basic uses for the Internet: mail, discussion groups and file transfers. E-mail is the person to person use of communications that resembles fax except that it carries no sending charges. This provides fast, cheap worldwide communication's ability to the average person. (Mail can also be sent to multiple recipients, but "broadcast" sending is known as "Spam," a highly disapproved of means of communication – in some cases illegal.)

Discussion groups (newsgroups) are on-line sites for message posting, debate, exchanging information and generally interacting with other users of that particular newsgroup. USENET a sub-net of the Internet has some 30,000 different newsgroups generating some seven million words of commentary per day. The total number of newsgroups on the Net is probably over 100,000.

File transfers are the primary means by which information is sent over the Net. These require direct Internet access (logging on). This process allows users to access other machines and copy their files. (Doing this without permission is called "Hacking," which now carries legal penalties.) The web sites found on the Internet are simply entranceways to the computer files located at the web site. When one accesses a web site one is actually downloading a file, which then appears on your computer as a web page displaying information. Web sites and applications that work with web sites are able to communicate because all web based content is encoded in a standard code (HyperText Mark-up Language, HTML). Some web browsers for example "Google" index more then a billion web pages. However, digitalization does not necessarily improve content.

Internet's Impact

One of the major effects of information technology has been the tremendous boost that it has given to decentralization. Centralized, totalitarian control has become more difficult as citizens gained more communications ability. Governments simply cannot maintain secrecy the way they used to and a host of activists of every type now record and disseminate information on every government deed and mis-deed. People are now finding out what is going on and finding out more quickly.

Although the Internet has a long history of political intervention, it resists institutionalization. The Internet continues to spread information in a way that resists control. Not that there is a shortages of those who would control it. Business wants an Internet organized as a market dedicated to private profit. Government wants the Internet to be more fully regulated to prevent challenges to its moral, economic and political agendas. The military correctly see the Internet as a strategic infrastructure and want it made more sabotage proof and secure. Others want the Net used to support democracy, even "excessive democracy."

Thomas Jefferson stated that free communications were the only effective guardian of the people's other rights. This viewpoint encourages the hopes of those who see intellectual freedom as a foundation for political freedom. However, accurate and timely information is also a key part of most economic systems, a process that makes information a valuable commodity and control of information a business activity.

Much of the real impact of the Internet and its associated technologies has been hidden by the dismissal of problems. Gains from increased efficiency and the effective elimination of distance are celebrated. Sacrifices in the form of lost jobs or lost capabilities are dismissed as trivial or simply ignored. The social changes associated with a massive restructuring of business and governments are touted as the most important changes of the decade, the century - or ever. The fact of losers, as well as winners is simply ignored. This does not make the gains unreal or prove that the gains do not outweigh the losses. Too often this offloading of the sacrifices simply accentuated the trend to a have and have-not society.

The increasing influence of commerce on the Net and growing government concerns over political impacts have not deterred those who seek a more Jeffersonian democracy. The goal of these activists is to make information flow not only more freely (that's happening anyway) but flow around power hierarchies and outside the control of the traditional information monopolies. This is seen as a new basis for citizen's power and as a method of restoring public participation in the political process. It is the desire of these activists (of all political stripes) to democratize Cyberspace and counter-act the overwhelmingly corporate control of the world's media.

The established model of central information control by government, media or publisher is being challenged by the explosive growth of the Internet. The top-down distribution of information is no longer the only game in town. Individuals now have the ability to set up web sites and "broadcast" their views, news, or arguments to large audiences. This forms a new basis for grassroots power. Net organizations have alerted and informed the world about student revolts in China and revolutions in Mexico. In Canada the successful opposition of the Multilateral Agreement on Investments (the MAI was a charter of corporate rights outside democratic control) was based on numerous organizations use of the Internet.

All points of view have become expressible and sharable on a worldwide scale. The Net allows a meeting of minds that can form and transform ideas. It bypasses the control of media owners and government censors. By doing so it combats the limited and biased points of view that are associated with limited facts and limited ways of understanding those facts. This access to a "marketplace" of ideas allows for more reasoned judgments at the same time that it increases the importance of the skills needed to discern accurate information.

This new marketplace for ideas and information allows like-minded people to organize outside the barriers of social and geographic isolation. New political alliances become possible, new forms of organization become feasible. The Internet also challenges the secure structures of our opinions and social consciousness by letting in a barrage of conflicting and contrary viewpoints.

Increasingly, the agenda framing and veto power of an information oligarchy is being challenged. The limitations on public discourse that were supported by establishing limits on ideas that could be expressed in a mass media are slowly being undercut by the emergence of the Internet as an uncontrolled mass media. As this occurs, the competition from uncontrolled voices challenges the elite's control of public perception. In response, the establishment has exerted great energy in trying to shape public opinion to support censorship. Pornography, the War on Drugs, "hate speech" and a number of other problems have been used to establish a pro-censorship agenda. These law-based attempts at censorship are supported by extensive efforts to establish "political correctness" as a socially based censorship.

The power to control information is more than the power to control what can be discussed or what facts become public. Controlling information also allows one to control conceptualization, to control the images and ideas that place us in relation to history and society. Information control allows the controller to structure the debate or prevent debate, to label what is significant and why.

The Internet breaks up this control by decentralizing the origin of messages, by resisting prior censorship and by allowing groups with few resources to be heard. It allows ideas and viewpoints to rise or fall on their merits rather than on their "acceptability." Increasingly, ordinary people are gaining the ability to form intentional communities in a participatory fashion. This imperils official reality and the political agendas that are justified by official reality.

The ability of the mass media to hide elite control behind outwardly democratic forms is slowly eroding. Political and cultural reality can no longer be programmed by an information oligarchy hiding behind a mask of democracy. It is unclear whether this will dis-empower the information establishment or simply lead to a progressive abandonment of the mask.

This situation raises serious questions about technical and educational barriers to the information commons. As propaganda, dis-information and agenda scholarship are used to preserve the power of established elites; the ability to see through them becomes crucial. Much of the power imbalance between the people and the institutions serving elite interests has been based on access to specialized information. Now, the Internet is opening areas of discourse that were previously reserved exclusively for the powerful. Meaningful and informative argument and debate are now taking place around constitutions, criminological and social issues that were simply "not discussed" even a few years ago.

The failure of mainstream journalism to address major areas of public concern and the narrowness of public debate are both under pressure from an increasing array of independent journalists and authors. Writers can now publish their work electronically on the Net, bypassing the approval of publishers and editorial boards.

Many activists and concerned citizens see these developments as a solution to the fragmentation and disempowerment of those fighting elite agendas. The ability of web sites to link to other like-minded sites increases their effectiveness. Electronic communities of discourse, political activism and elite criticism are a growing reality on the Internet. How these communities will find the funding to support alternative scholarship, journalism and newsgathering is still unclear. That powerful groups with large resources would like them to fail is all too clear.

Many activists and critics of the system worry that censorship established for pornography will evolve into political censorship. This belief is firmly rooted in the loose wording and uncertain scope of some of these laws. This "morality agenda" is being expanded to include censorship in the form of "drug information" laws and "hate speech" laws often with very loose and expandable wording. This censorship creep is already inspiring some Internet providers and gateway sites to reject sites that are "too controversial" or "too political."

Corporate ownership of the fiber optic networks that support Internet traffic also threatens continued access to the Internet. The possibility of censorship by "commercial viability" or by the mandating of high cost technologies is also a concern for many. The Internet may be "owned" by everybody and may have been developed by public money, but ownership of the cable and the networking computers is in private hands. There is currently very little in the way of "common carrier" legislation to protect access.

Aldous Huxley's dark vision of the future was a future based on intellectual tyranny. The people would serve their masters happily because they had been convinced that servitude was best or - even better - that their servitude was not servitude at all. To make people accept their "place" and their "lot" is the objective of elite information control. To prevent meaningful questioning of "things as they are" or the "natural order of things" is ever the activity of those who profit by the status quo. However, making people love their present condition is best and an age-old agenda for "Ministries of Truth," "Official Religions," news managers and schoolteachers.

It is unclear how the democratic challenge of the Net will survive as it is increasingly commercialized. As the first new mass medium in 50 years, it's currently challenging the agenda setters. How or if the struggle to maintain the Net as a democratic mass medium will be won is unclear. The stakes are the future itself and the power to frame that future.

Knowledge and InfoWar

To understand the ongoing struggle to control information and shape ideas we must examine the nature of knowledge itself. Unfortunately, "knowledge" is a vast subject that philosophers have disputed over for centuries. Nor have the various academic disciplines been able to find a simple and uncontroversial approach to knowledge. This leaves us with a bit of a problem. If we simply skip the discussion, we will hinder our understanding. If we attempt a complete discussion, we will attempt what two thousand years of philosophy and debate have failed to resolve!

This chapter will attempt a middle course. First, it will describe and discuss some basic approaches to understanding what "knowledge" is. Second, it will examine knowledge as a social artefact. In the course of this intellectual journey, we will gain a better understanding of why ideas are so important. If we understand why ideas are important, we will also understand why they are so contested. Why they shape the InfoWar now raging in Cyberspace (and elsewhere). In an era of dis-information and agenda scholarship, these understandings are a necessary tool for any researcher. Therefore, with apologies to philosophers, meta-physicians and other savants for my chopping and dicing of the fine points, I shall begin.

Some Basics

Readers are sometimes confused, or angered, by the assumptions that various authors make. In some cases, these "taken as givens" seem to make no sense. In part, this is caused by different authors having different philosophical approaches to knowledge. Simply put, knowledge can be approached from "Positivist," "Realist" or "Relativist" assumptions.

(Note: I'm putting words in quotation marks to alert the reader that these words are being used in a technical sense with meanings that are more specific than their meaning in everyday speech. For example, a "religious" argument means an argument based on faith, it does not imply Religion in the spiritual sense of "a religion.")

Positivism, as the name implies, is a belief in our ability to identify "absolute" or "positive truth." In this view things are what they are, - solid, unchanging – unaffected by our actions or intents. For Positivism, the "truth" is our there. This has the advantage of making things simple. Once "the truth" has been found then no further discussion is necessary. Unfortunately, Positivism has been unable to answer a critical question. "How do we know it's true?" There are two possible answers to this question. You know it's true because you have a perfect test to prove it's true or you have the knowledge from a perfect source - an argument of authority.

Unfortunately, the perfect test approach (for example Popper's falsification criteria) has failed because you must first establish that your perfect test is perfect. So using the "perfect test" solution merely shifts the problem from finding a perfect "answer" to finding a perfect "test." There is also the problem of determining how to apply our "perfect test" perfectly. No one has been able to find a solution to these problems – or at least been able to prove that they have successfully found the solution. (If you're getting the impression that grabbing hold of "the truth" is a little slippery, you're getting the idea.)

The second "proof" – an argument of authority - is also simple. If the chosen authority says it's true then it's true – end of discussion. Unfortunately, this argument is based on faith not reason. If you have "faith" in the authority – Adam Smith, the Gospels, the "philosophers," whatever – then all is well. However, there is no way to prove that the authority actually is right (except by "faith"). This "proof" to Positivism is simply a retreat into "religious" faith.

It is now uncommon to find "hard core" Positivist arguments outside of religion. However, people often bring Positivist elements into their arguments. For example, "economic forces" or "invisible hands" might be asserted as the cause of something, an argument that establishes them as forces of nature. This approach gives the impression of certainty, but that certainty is a matter of faith in the authority not fact. (There has been a tendency to regard some social sciences as if they were natural sciences with "natural laws" akin to the law of gravity. In essence, this is a Positivist viewpoint.)

Relativism, as the name implies, is the belief that things are "relative to the beholder." This approach, sometimes called "Post-modernism," works from the understanding that we cannot "prove" anything in the absolute sense. Since we cannot prove anything, the only approach available is to take the eye of the beholder as our reference point. This approach (sometimes called Subjective Relativism) asserts that things are whatever they appear to be. The "proof" that they are what they appear to be is that the beholder believes it!

A classic example of a "hard core" Relativist viewpoint is the story of natives who are seeing a sailboat for the first time and never having seen a sail before they imagine that the boat is an unusual cloud on the horizon. A Relativist would say that they are right; if they "believe" it's a cloud then it is a cloud. (At least to them, to us it would be a boat – both views would be equally "correct.")

There are some obvious problems with this approach. If anything can be anything and there is no way to determine if one opinion is better than another, then all intellectual activity collapses. This is a system where only "belief" is possible not "knowledge." Detaching "belief" from "objective reality" in this manner can lead to some amazingly complex and interesting intellectual games and speculations. Unfortunately, it's not terribly useful in sorting out the "everyday world."

Relativist arguments are usually encountered in a watered down form, often in assertions that "nothing can be known for sure" or in the belief that all opinions are of equal worth. Relativist arguments are sometimes used to assert "equality" as in "everybody's opinions are equally valid and worthy." Relativist arguments, or assumptions, are also used to keep investigation or discussion going indefinitely - thereby preventing action.

Realism is a mixture of Positivism and Relativism. It is "Positive" in the sense that it believes in an "objective reality," that is a reality that is independent of our intents or perceptions. (A sail is a sail not a cloud, if you think it's a cloud you're mistaken.) However, it is also "Relativist" in its belief that we "filter" our perceptions through our ideas, beliefs and prejudices. This usually leads to a "correspondence" theory of "truth." If what we belief corresponds to what "really is" then what we believe is true.

From this viewpoint, there is no "absolute truth" in the Positivist sense (or if there is, we cannot be sure that we "know" it). However, there is probable truth. Things are always open to question but over time inquiry, analysis and investigation determine that some things are "probably true." This tends to form a continuum from wild speculation to virtual certainty.

The Enlightenment's creation of modern science is founded on a Realist viewpoint. Everything is open to question, there are no "givens," nor is anything ever absolutely or finally settled. However, in practice some things are supported by such overwhelming evidence that they can be taken as "true" in the practical sense. This view asserts the "solidness" of what is and lays great emphasis on inquiry. However, it also recognizes a human or social element where our beliefs, desires, agendas, and ideas can "shape" what we "see." Since our ideas about how the world works shape how we evaluate and understand "objective" reality shaping a person's ideas will also, to a certain extent, shape their viewpoint. Accordingly, if you can shape a person's ideas by controlling the available "facts" you can influence how they see things and how they interpret what they see. This social element, that ideas shape perception and understanding, leads to the political realization that controlling the ideas and "facts" available to a person is an excellent way to effectively control that person and their actions.

Knowledge as a Social Structure

Since our ideas, viewpoints and agendas can shape our perceptions, we can conclude that all knowledge, or at least knowledge available to human beings, has a certain "social" element. This human element in thought is a "social" element because the ideas, viewpoints and assumptions that underlie our thinking are not individually constructed solely from observation. As individuals, we inherit these ideas and arguments as well as receiving them from our contemporaries. As the sociologist Karl Mannheim first stated in the 1920's, knowledge is socially connected and socially structured. Knowledge, in short, is a social/ historical artefact tied to experience and particular knowledge is attached to particular social/historical groups. Since so much of the struggle on the Internet is a struggle between different groups with different agendas and different viewpoints, it's useful to examine knowledge as a social artefact. In short, to look at how people really think.

Multiculturalism and globalization have heightened our awareness that there are different styles (ways) of thinking as well as different contents (facts, theories) of thinking. Our realization that there are systems which have conflicting assumptions, concepts, values and procedures opens the possibility of addressing knowledge from a systems approach rather than from an information flow or contents approach. The insight that our ideas effect our perceptions can be traced back to Francis Bacon's "idols." Bacon traced intellectual errors to human nature (idols of the tribe), individual biases (idols of the cave), social influences (idols of the market place) and false philosophies (idols of the theatre).

This understanding that our ideas, and the ideas of the social groups we belong to, affect our perceptions of the world underlines the realization that knowledge is a social artefact and emphasizes the constructed nature of knowledge. Much of what we think and know is based on what we have been taught or given. Indeed, all knowledge is in part, historically driven. Whether our knowledge comes from the Bible, Adam Smith, or Thomas Jefferson we are "thinking further" from a historical accumulation of ideas, facts and insights. (This is one of the reasons that people into thought control work hard to control how "history" is written and interpreted.)

This Realist approach to knowledge draws distinctions between the natural and social sciences. A separation based on the perception that natural and social science disciplines are dealing with different types of knowledge. The natural sciences deal (largely) with the non-human world, a "world" that is largely disconnected from human meanings and values. This "justifies" an analysis using a more "absolute" kind of "truth." However, social knowledge, knowledge of the human world, is innately tied to meanings and values, a process that adds a social/historical dimension to the concept of "truth." This insight conforms to the Realist belief that the certainty or reliability of knowledge is located along a continuum from speculation to near certainty. For example, knowledge such as the "law of gravity" has been well tested, is easily observable and hence is very certain. Knowledge such as "welfare payments cause poverty" is tied to a host of assumptions, subject to problems in measurement and tied to a number of beliefs about the nature of "welfare" and "poverty." Hence it is much less certain and more debatable. Some knowledge for example, "marijuana is bad" is so tied to qualitative value judgments that it is extremely difficult to be certain in the empirically sense. (Ethically certain is another matter but ethics are in the realm of social, cultural "belief" often based on "faith" which is a different "order" of knowledge.)

Because of the historical driven nature of knowledge, different groups often attach different meanings to the same object. For example, some people see guns as a symbol of violence; some see them as a symbol of freedom. These different viewpoints are often tied to different personal or group histories. A group of people in a housing project who experience guns as - something only gang members have - are likely to view them quite differently than a rural group with a long tradition of sports shooting and hunting. Neither group is "wrong" in the literal sense. Both are viewing the object from their own, quite valid, experience. Objects are seen differently depending on the viewpoint of the observer. This viewpoint is often historically driven.

These facts demonstrate that there is a relationship between knowledge (in this case what they "know" about guns) and viewpoint. Knowledge is socially conditioned and recognizing this relationship opens the possibility of understanding the process. An examination of the existential, social underpinnings of ideas and beliefs also

helps restore communication between social groups
Understanding the social basis of different viewpoints
allows people to "translate" across different perspectives
and overcome the problem of people "talking past" one
another. (We can note here that overcoming the "talking
past one another" problem is essential for consensus and
alliance building. People who want to prevent alliances
or consensuses building often strive to prevent this
social/ historical understanding of knowledge.)

(Author's note: I once saw an example of this on the
Oprah show. An audience was discussing abortion and
as one would expect, there were varieties of pro and con
opinions. However, as the debate went on there was a
growing separation between the black and white
members of the audience. This separation quickly
developed into a "what's the matter with you! Only an
Idiot would look at it like that!" fight. This happened
because both groups had an underlying "common sense,
everybody knows that" viewpoint that shaped how they
were approaching the issue. The black members of the
audience "simply assumed" that abortion was an issue of
preventing birth. The white members of the audience
"simply assumed" that abortion was an issue of enforcing
birth. What was not recognized was the historical nature
of this group conflict. Blacks in the US have been on the
receiving end of government policies to discourage the
growth of the black population. Whites have been on
the receiving end of policies to do the exact opposite.
Each group had a, perfectly true and valid, historical
experience that structured how they "just naturally" saw
the issue. The point of commonality (never reached
because no one saw the source of the disagreement) was
enforced government fertility policies – in this case
applied differently to different groups. The underlying
debate over government fertility policies simply never
happened.)

Different historical and social experiences lead to different understandings of the world. Moreover, different social groups, by the process of repeatedly dealing with typical existential problems, developed particular modes of thought and types of knowledge. Consequently, the structure of thought - as well as the content of thought - differs for different social and historical groups. This means that the same object will appear differently to people with different social relationships to that object. This phenomenon, the social variability of knowledge, raises the question of when and where social structures show up in the structure of assertions and in what sense social structures determine knowledge. These socially constructed viewpoints can be called "perspectives," defined as a person's whole mode of conceiving things. Perspectives are sometime referred to as worldviews. (We can note that some perspectives or worldviews support the status quo while others challenge it. Identifying the "politics" of a perspective gives a lot of insight into an author's agenda.)

This "social artefact" approach to knowledge and analysis can be used as a method of empirical research. As a method for purely empirical investigation, it uses description and structural analysis to examine how social relations influenced thought. Through this process, one comes to understand different styles of thought as a reflection of group (or individual) viewpoints. This theory of knowledge can be used to explain how social/existential factors influence the construction of knowledge and shape the development of criterions of validity. This approach stresses the "intentional" and "active' character of knowledge. An example of this in action is the checks and balances approach of the American Constitution. This approach was chosen because their "knowledge" regarding the "best form of government" was developed as a response to the founding fathers historical experience with the "over-powerful monarch" George III, hence the desire for a system that dispersed power.

Knowledge as a social artefact is not just a matter of content – what is known. A group's knowledge also has a theoretical structure. The categories, concepts and assumptions of that structure shape the information that it contains. These concepts, the theoretical structure, are an intellectual response to the group's typical problems and goals. This theoretical structure is also shaped by the group's interests and its competitions with other group. The historical nature of a group's viewpoint ties that viewpoint to its experience in the world. This structure, with its assumptions and "common sense" viewpoints, holds and shapes the particulars of the group's knowledge.

This social genesis of knowledge in turn implies that human will, especially as expressed by groups, cannot be separated from knowledge. Every formulation of a problem is based on a human act (the question shapes the answer). Every formulation of a problem is guided by a group's experience with similar problems. In every selection of data, there is an act of will on the part of the knower. The object "in itself" determines thought, but thought is also determined by the different expectations, purposes and experiences that the group (or individual) has with that object.

People use their brains and their intellects to deal with life-situations. Consequently, the problem of "knowing" becomes more intelligible if we understand that our "knowing" is connect to our (and our groups) life in the world. The only kind of knowing that is available to us (baring "religious" knowing) is knowing as a human activity, as a human act. Human knowledge, especially group knowledge, is shaped by how the group deals with its life situations and by its social and historical location. It is true that the single "individual" thinks. However, it is also true that they participate in thinking further what other people have thought before him. The individual finds themselves in an inherited situation with patterns of thought, which are more or less appropriate to their situation, and attempts to "think further."

This group process of "thinking further" does not imply a group mind or a collective consciousness. It refers to a socialization of meanings that shapes the perceptions of group members. The inherited categories, concepts and modes of thought of a group stabilized its identity and shaped its member's consciousness. The individual is the bearer of knowledge, but this knowledge is conditioned by the group's collective understandings. These collective understandings, expressed in the structure and contents of the group's knowledge, are disseminated through the process of socialization and communication.

This group perspective is not something that the individual is forced to accept. However, a person's perspective is necessarily chosen from the perspectives that they know about or are prepared to construct on their own. (This is one of the major reasons for censorship.) In the intellectual sense, a person doesn't belong to a group because they were born into it, or because they claim to belong to it. Nor is group membership simply a matter of political loyalty or social allegiances. An individual belongs to a group (in the intellectual sense) because they see the world in terms of the group's meanings and viewpoints. In the intellectual sense it is our choice of meanings that establishes our group memberships.

The conceptual systems developed by each group are a consequence of active economic, political and social competition. Their viewpoints are ultimately the intellectual expression of the conflicting group's struggles for power. Attempts to persuade, "educate" or socialize people into a group's perspective are political activities aimed at gaining the group strength needed to achieve political goals.

This "socialization of thought" is not uniform. Some thought, such as the type of thought exemplified by the expression 2x2=4, are "socially detached." However, the more closely the thought is to the subject matter of the social sciences, the more closely it's connected to group experiences and agendas. A person's social and political perspectives contain "wilful" elements. These "wilful" elements have a "qualitative" nature that is an unavoidable consequence of the pragmatic nature of people's thinking and people's habit of organizing their thoughts according to their interests. From different perspectives, "the object" has different meanings because people's perception of the object is tied to their respective frames of reference. This "qualitative" factor, the presence of "meaningful" elements in the social perception of objects, is a basic characteristic of all social analysis. This element of meaningfulness in thought is a source of analytical insight and a problem. The presence of "meaningful" elements in social knowledge makes the analysis of various interpretations problematic.

This understanding of knowledge raises analytical problems because there is always (implicitly or explicitly) a model of how thinking can and should be carried out. This understanding of social thought can be addressed with the concept of relationism. Relationism asserts that all thought has social roots and is "related" to human interests.

In order to understand the social roots of thought it is necessary to analyze the concrete, empirical connections of specific ideas to specific social groups. This exercise makes it possible to draw an intellectual map of society, where a specific group's perspective would constitute a cultural and intellectual index of its historical/social position. For a researcher, noting how a particular word or concept means different things to different groups is often the key to understanding group differences. The observer's interests determine how concepts are formulated. The absence of a concept, the absence of certain points of view, can also indicate the absence of a desire or need to come to grips with certain issues.

Analyzing the concepts in a given conceptual scheme provides the most direct approach to the perspective of distinctly situated group. A researcher can identify a perspective by the traits and interests that locate it in relation to a situation or an era. A perspective can be defined by an analysis of the meaning of its concepts, by the phenomenon of the counter-concept and by the absence of concepts. It can also be identified by the structure of categorical apparatus, by the dominant mode of thought, by characteristic levels of abstraction and by what elements are included in the perspective.

The understanding that a perspective (or worldview) is tied it to a group's social position raises an analytical problem. If all though is social and tied to group interests how can one be objective? If we try to answer this from a Positivist perspective based on the ideal of an absolute (but unverifiable) truth we find that we have reached an impasse. However, if we acknowledge the limits of human knowledge, we are led back to the

Realist perception that knowledge necessarily contains a human element. From this Realist perspective we can compare different viewpoints to enlarge our own viewpoint and engage in the analytical process of determining what is the most "probable truth."

This movement towards a more inclusive perspective is supported by the belief that each group's perspective is particular to that group and consequently only a partial perspective. This wider view also tends to be more abstract and use categories that are more general. This analytical process of Perspectivism allows two approaches to objectivity. Objectivity can be studied within the boundaries of a single perspective or objectivity can be studied across the boundaries of multiple perspectives. Observers within a single system share an identical conceptual and categorical apparatus and would participate in a common universe of discourse. Accordingly, their analysis would tend towards similar results, and common criteria could be used to eliminate "error."

However, if the observers have different perspectives they would have to arrive at objectivity in a more roundabout fashion. Differences in perception would have to be understood in light of the differences in structure of the different systems. Therefore, an effort would have to be made to find a formula for translating the results of one system into those of the other and to discover common denominators for these varying perspectives. In practice, resolving differences across perspectives is addressed by giving precedence to the perspective that is most inclusive. That is, analytical precedence is given to the perspective that is closer to a "totality" of vision, as well as perspectives that have a greater analytical "utility."

This approach is based on the understanding that there is a relationship between social existence and analysis. Analysis cannot be separated from the social determination of meaning and the social determination of validity. However, this approach is dangerously close to Relativism. It comes close to saying that social determinations (beliefs) determine validity. The saving feature that prevents this collapse into Relativism is the concept of Relationism, that social beliefs have a "relationship" to the non- social world, that the "real world" of the natural sciences, of the actual physical event "grounds" the analysis. The idea of Relationism, that we are "grounded" in the real physical world, gives an analytical basis for determining "the truth" or at least "probable truth." In a sense we can have "true" human knowledge because we are not separate from the "real" world. Despite our agendas and illusions, or hopes and our manoeuvring the real world eventually "pushes back." The actual, empirically verifiable, physical fact establishes analytical stability and saves us from unlimited speculation and spin doctoring. (Relationism has two aspects: a relationship between ideas and our "perceptions," and a relationship between our "perceptions" and the "real world.")

This approach is also separates different types of "truth" for different "spheres" of reality. There is the "truth" associated with the physical world of the natural sciences and little affected by our imaginings. There is also the "truth" of our social and cultural worlds, strongly affected by our intellectual efforts. Some forms of knowledge are not tied to universal facts of nature but are instead tied to culture. Therefore, the "truth" of culture, the world of human meanings, is a historical variable assessable through the experience of culture. Relationism, in the middle ground between Positivism and Relativism, ties "truth" to concrete historical/social situations thereby allowing criteria that are if not absolute are at least finite and identifiable, hence usable. (When we say something is "true" from an "economic perspective" or "true" from an "ethical perspective" or for that matter "true" from a "strict Constitutional perspective" we are acknowledging the variability of "cultural or social truth.")

Analysis is not dependent on having an "absolute sphere of truth," "in itself." It is dependent (at least in the social/ historical world) on the conceptual structure of knowledge in each historical period. Therefore, each perspective's claims to validity and its criteria of validity are socially/historically conditioned like those of any other perspective. This approach ties "truth" or "valid analysis" to historical reality. Objectivity is achieved by making comparisons and attempting a more inclusive analysis.

This social or cultural factor in analysis supports a comparative methodology. In this approach, different perspectives are seen in terms of their different purposes, motives and intellectual structures. Therefore, the existence of different views is not necessarily an indication of the existence of errors. By cross checking between perspectives the analyst can identify common denominators and isolate extraneous, arbitrary elements. The analyst's perspective becomes more inclusive by becoming more abstract, by dissolving the earlier more particular and concrete points of view into a more abstract formulation. In the wider perspective, the qualitative, historical particulars of concrete perspectives were absorbed into the general and abstract.

This social determination of viewpoint also affects the analyst. The analyst, being part of the human world and being a member of a group, is within the matrix of history and culture. As such, they are not neutral, detached observers. This raises questions about the analyst's ability to analyze freely. This is a conflict between the social determination of knowledge and the analyst's free will. From a determinist viewpoint, we can note that their history and social position shapes both humans and their intellectual tools. From the viewpoint of free will we can note that there is always a choice, even if a historically constrained choice, in our perception of the world. Our human involvement determines that our viewpoint is based on our cultural values and beliefs. This makes our social world, to an extent, a self-fulfilling prophecy. Yet there is always a choice of meanings and values, and therefore always a choice in how we shaped our world.

Both "facts" and "values" rise out of life experiences; there is no social fact without valuation and no valuation without some factual context. Human values are inseparable from human interests and human concerns. For most societies, the basic motor that determined events is competition. In the final analysis, all social groups are shaped by human interaction with the material world and by the human interactions of the social world.

In sorting out the complex analytical puzzle of these interactions an awareness of the social determination of knowledge is the key and, in a sense, the antidote to subjectivism and bias. In this approach to knowledge, the category of "absolute" truth is replaced with a more realistic category of "probable" truth. Analysis is moved from the "separate" sphere of the "unattached observer" to a social/ historical sphere of knowledge.

Outside of religion, practical knowledge means human knowledge; human knowledge that is and will always remain, at least in part, subjective. For human knowledge cannot be separated from its human bearer, its human element, and still by definition remain as human knowledge. Analysis, especially in the social sphere, is a human centered activity, were an awareness of our social/ historical foundations can inform us of the pitfalls of our social/ historical viewpoints.

Computerization and Thinking

The Internet is technologically dependent on the computer. The Net comes to us through our computers, sharing the prestige that we attached to computers, computerization and all things digital. However, the computerization of knowledge is a complex process that has brought a number of assumptions and assertions with it. This chapter examines the idea that computerized data (or information) is the same thing as "knowledge." It also examines the concept of the "thinking machine" and its relationship to "knowledge." Those who would use the global communications network for research need to understand the issues raised by a computerized approach to knowledge.

Packaging Knowledge – Packaging Thinking

Computers make us smarter. Computers make us more efficient, more effective, more competitive and more productive. Even the mere presence of a computer – in every classroom – improves education. In an era of dissatisfaction with educational standards and the dumbing down of society, computers are the answer to every problem. Computers will re-invent business, increase our leisure time, eliminate waste and lead to the "paperless" office. Computers will ……..

Had enough yet?

For at least a decade and arguably for two decades we've been hearing this techno-babble about the wonders of computerization. Curiously, as we spend billions of dollars on computers and on computerization things seem to be getting worse. We have less leisure, society continues to "dumb down," infotainment replaces news and our offices are busier and more snowed under with paper than ever.

Now broad band Internet connections will, we are told, finally usher in the new computer age and allow us the true multimedia capabilities of on-line movies and interactive video. Some computer advocates, for example Prof. Papert of MIT have even argued that interactive multimedia software has become so advanced that "point and click" can replace reading. Experts assert that the modern computer, supported by "expert software" and "user friendly" interfaces, has placed us within reach of a "knowledge machine." This machine or series of machines will spare us the expense and effort of acquiring difficult intellectual skills like literacy and numeracy.

The idea of the "knowledge machine" is not new either in science fiction or in computer advertising. The promise (or threat) is of a machine so powerful, with such a large store of data that it can order, arrange and process information for us. This information will be so arranged that even reading (beyond a very basic level) will not be required. A simple point and click picture interface will lead the seeker to the data, information, or knowledge that they seek. The Internet with its ever-increasing store of data and information is slated to be the "data bank" that will house libraries full of information. This data will be available "on call" by the user or for that matter, by the "intelligent machine." The computer and the Internet will combine (the integration of the various media is being referred to as "convergence") to provide easier access to greater and deeper bodies of knowledge.

Unfortunately, there are some problems with the brave new world of unlimited information via "smart machines." The first problem is that it presupposes something that does not exist. The proponents of the "knowledge machine" imagine that the machine "makes" the information. This is incorrect. People "make" the information just as people made the machine and the software. The dream of removing human error and human bias by shifting the decision making to machines is just that - a dream. In reality, this process merely shifts the problem of decision making back one step to software designers and data technicians

The dream of a "knowledge machine" usually skips around the issue of who will program and control the machine. The increasing commercialization of knowledge, seen in the expansion of intellectual property laws and the increasing span of what can be copyrighted, directly effects how a "knowledge machine" would be organized. The increasing commodification of knowledge as a marketable commodity or as a competitive advantage will also influence any attempt at a "knowledge machine." Further, there is the whole political question of who will be the information gatekeepers. As we have discussed in the previous chapter the control of how knowledge is organized and how it is presented are important political questions. The failure of the recent attempts by the US government to develop a "standard" curriculum is an indication of how politically sensitive control of knowledge can be, especially if that knowledge is tied to the determination of an "official viewpoint" or an "official interpretation." The Federal government's attempt to establish a national history curriculum with national "standards" foundered on this political rock.

The idea of a computer based "knowledge machine" also raises profound political questions. One of the most important is the whole issue of reading. While we tend to skip over this issue, unless were discussing "functional literacy"; reading has tremendous political importance. A person who cannot read, or who only reads at a simplistic level is intellectually handicapped and very dependent on being told what they "need to know." It's relevant that slaves were once forbidden to learn how to read. Nor is it coincidental that there was strong elite opposition to the working class being taught to read beyond the barely functional level.

Reading is the primary means by which we acquire and process information. The printed page quite literally opens whole new worlds for the reader. Books and the ready availability of books bypass the filtered information of the mass media. It also bypasses official information control. The ever-increasing assortment of reading material available on the Net, from every conceivable source, is a fact of major political importance. In a world with increasing government censorship and thought control; in a world were the vast majority of all mass distributed information is tied to corporate financial interests, reading has become a political act.

Efforts or agendas to "replace reading" with an effortless "just click" approach to knowledge must be seen in this context. The widespread agenda to establish literacy standards at "functional literacy" must also be seen as a profoundly political act. (This is often accompanied with the not so hidden rational that that is all that "these people" are capable of.) To go along with this agenda is to effectively separate large numbers of people from alternate sources of information and make them dependent on those in control of the "machine." It is not difficult to imagine which groups will establish the "correct viewpoints," the "right answers" and the "relevant facts" for any pre-programmed "knowledge machine."

Reading is also basic to the development of the imagination. It stimulates the creation of our own personal vision, visions that necessarily vary from the "too exact" descriptions of multi-media. Consequently, these personal images are a source of diversity and creativity. Through them, we not "they," imagine.

This pre-packaged approach to information and knowledge has already had an impact on education. There are interest groups who are now calling for a "computer literacy" (whatever that might mean?) to supplement or even replace literacy and numeracy. This increasing reliance on the "knowledge machine," the "computer in every classroom," raises questions about whether students are being taught to think for themselves, or if priority is being given to "content" and "marketable skills."

This raises questions about the desirability of a computer driven curriculum, especially a curriculum that places increasing importance on "information" and the testing of "information" with a decreasing interest in critical thinking. The shift to statewide multiple-choice exams reinforces this "information approach" to testing. It establishes "standards" that by their very nature lead to the testing of disjointed "factoids." Analysis, reasoning and critical thinking skills can not be tested with multiple choice questions. We can observe that this approach to education serves the - information = knowledge – model of "knowledge" that is so in tune with the computer's abilities. This model is also in tune with a distinctly top down approach to knowledge.

This approach to "knowledge," both in and out of schools, blurs the distinction between education and training. To understand the difference one must understand the process of learning. Learning is based on three things: acquiring information, processing information and contents (the information). Acquiring information is a matter of literacy and numeracy. Processing information is a matter of reasoning, critical thinking and analysis. Content, is the information itself.

While both training and education require "contents," education stresses the first two elements "collecting" and "processing." Training stresses contents. If a person is "educated," they have developed an ability to find and process information (literacy, numeracy and critical thinking). With these skills, they are free to "learn" whatever they want. They can acquire knowledge independently, in conformity with their own agendas. A person who has been "trained" is dependent on others to determine what they "should know" and to present that information to them. This is the essential difference between "education" and "training." Educated people are free to develop their own knowledge. A trained person's knowledge is dependent on others. (The actual level of knowledge or skill is not the issue. A marine, for example, may be highly skilled and knowledgeable about his trade but he is not "educated.") Clearly, the determination of "education" vs. "training" is a serious political question.

The advocates of a new "knowledge economy" which is part of the new "information age" are relying heavily on "computerization" to supply expertise and knowledge. There have been widespread instances of computers leading to the deskilling of the workforce. Expert systems (systems with built in "decision making" ability) are being used to replace managers and skilled professionals. The computer's ability to store information and the processing power of software are central to this agenda. However, it envisions data storage as "memory" and information processing as "thinking." Critics have seen this process as the replacement of imagination and reasoning with low grade counterfeits.

Software and Knowledge

It is questionable whether a "knowledge economy" and an "information age," are really Utopias. Technological change, particularly the computer/ telecommunications revolution of the Internet, has problems as well as advantages. One of those problems is the technology tends to obscure the power and political arrangements that support it. The Internet has picked up the technological optimism that was once associated with computers. It's the latest "silver bullet" to deal with society's political, economic and social problems. Computers and the Internet do this in part because they blur the distinction between data, information and knowledge. They also blur the distinction between information, imagination and insight.

The computerized approach to information assumes that information is a homogenous category, that there are no qualitative differences between different bits of data or information. From the computer's viewpoint Shakespeare is the same a stock quotes, falsehoods are the same as truth. It's all just data. However, in the real world the qualitative aspects of information are important. Information is important in relationship to questions – that the information can or cannot answer. Information is also important in its relationship to human needs and desires.

The ambitious project of the "superbrain" has fallen prey to these sorts of problems. Computers can concentrate "decision making" powers but they do so through a technological filter. "Expert software" and problem solving software promise "intelligence" but in reality they are simply cookbook pro-formas based on the programming team's decisions. A program that delivers "expert advice" on any subject is in reality just a logic tree made up of the insights of people familiar with that area. Programmers simply ask the experts – if this happens then what? If this situation is present then what? These questions and answers are built into the software. The machine then displays them as the user triggers the selected response. Complex and sophisticated as these programs are, in essence they remain nothing more than computerized opinions. They are limited to the questions and answers of their makers.

Computers are linear devices. They exist in a world of "logical" cause and effect. They are very Cartesian in their rejection of intuition, inspiration, serendipity, and circumstance. They de-contextualize information in abstract arrangements that ultimately are tied to the mathematical models that form the core of software design. The computer can only perform tasks that are reduced to a series of discrete unambiguous steps. A computer has no "judgment." (Technically, software design is based on algorithms. An algorithm is a sequence of precise instructions telling the computer what to do. Computers do not "think" in any meaningful sense of the word. They follow their software instructions, which can give the appearance of "thinking.")

Computers can be very useful tools for processing large amounts of data. They rescue us from a lot of merely mechanical work. They do not tell us what data to process, or how to process it or what to do with it once it is processed. (They can "tell us" by repeating what the programmer told them to tell us.)

As computers do this processing, they concentrate information control. They do this by giving decision-making authority to the programmer who "builds" decisions into the software. They also concentrate control by giving greater information processing power to the computer user – or the employer of the computer user.

This increasing in processing power has led to more information processing and greater prestige for information processing. In the same manner that easy to use statistical programs increased the use and prestige of statistics, our shiny new ability has given us the desire to use our new abilities. The traditional view of information, that it's simply a category holding disjointed matters of fact has been replaced by a "cult" of information. Processing itself has become a prestigious billion-dollar industry.

Business and government are putting ever-greater resources into data management. The process of collecting, manipulating and publicizing data is leading to an increasing bureaucratization of life. More things are regulated and tracked because more things can be regulated and tracked. As this happens the white-collar professions are being subjected to the same automation and mechanization pressures that blue and pink collar workers have suffered through. As this trend advanced management in both business and government has pursued the dream of radically downsizing their workforce and/or radically deskilling the majority of jobs. The dream of the "thinking machine" has spurred their efforts at the same time that it has intimidated their workers. When the "thinking machine" has failed to handle qualitative data or exercise judgment the response has often been the devaluation of qualitative factors. The vastly complex process of human thinking has been cheapened by its reduction to an "information processing" model.

Following Claude Shannon's ground breaking work in communications theory ("A Mathematical Theory of Communications," 1949); communication has increasingly been seen as a quantitative event. The semantic contents of "meaning," have tended to be replaced by a quantitative "calculus" of communications based on quantities. This shift is so subjective in nature that the word "information" can be used for almost anything. It has become an all-purpose weasel word tied to data flow and separated from data content. This shift to a computer standard of "thinking" has tended to obscure the political, economic and social interests that have benefited from these technological changes.

The New Economy and "Thinking Machines"

The belief that information is "neutral" has helped to keep the political agendas of the "information age" hidden. Indeed, since the 1970's we have seen an extensive marketing campaign to generate support for the shift from smokestack to information industries. This shift has been very popular across a wide range of political opinion. For those on the Right of the political spectrum the shifting of large segments of the nation's industrial base to low wage countries has been masked by the argument that these are "old fashioned" industries of little worth. It has even been sold as a move to a "post-industrial" era. Why this "post-industrial" era only appears in high wage areas and not in the low wage "free enterprise zones" of the world goes unexplained.

For the Left of the political spectrum the shift to hi-tech has been seen as a positive ecological good as we move to more "non-polluting" industries. Electronics, aerospace and telecommunications are all glamorous and new, almost by definition "progress." This shift away from the physical to more "intellectual" industries has been mirrored and supported by a shift from running companies for profit to fiscal manipulation and treating companies as commodities. This is not to say that there have not been gains, winners, or indeed real progress. However, much of this political, economic and social re-alignment has been presented as a new "manifest destiny" agenda that in practice short-circuited political debate and co-opted rational policy making.

The "information economy" is based on forty years of government promotion and billions of dollars of public investment. There are few areas of the current hi-tech sector, which are not based on government investments. The rise of these industries has been accompanied by the domestic retirement of two generations of industrial capital and its re-investment overseas. This industrial shift has profoundly affected the people of the country. The domestic decline of the "old" smokestack industries has undercut the middle classes that based their prosperity on industrially skills. The downsizing of the "industrial middle class" has been so deep that even the idea of an "industrial middle class" is being challenged as "outmoded."

Much of the new hi-tech economy is a two-tier economy with a small and highly skilled elite and a much larger sector of contract and temporary workers often at low wages. Hi-tech also separates into domestic components (often at high wages) and foreign components, where manufacturing is concentrated at low wages. When one examines the rhetoric of the "experts" who applaud and support this agenda, it becomes clear that many of them have strong financial interests in an agenda sold as "best for everyone."

This agenda and the rhetoric that supports it have led to some very distorted ideas about computers, the process of thinking and how research is done. The "super machine" myth has led to the assertion that the central event in "thinking" is data processing. This leads to the downgrading of human thinking by reducing it to machine terms. It has also been a process of humanizing the machine. The vocabulary tells the story: "memory," "generations" of software and hardware, "evolving," etc. With this linguistic slight of hand, human or social evolution is tied to progress in machine performance. The social complexity of computation, communications and their various control devices is softened and simplified by this humanization – by implying that this technological development is somehow a human evolution and not just an improvement in tools. In our love affair with the machine, computer processes have become our prevalent models for "thinking."

This has been accentuated by the almost unlimited expansion of information access. A blizzard of available, sometimes too available, information has made us eager for any method of coping, even if that method blurs the distinction between raw data, information, knowledge and judgment. Since the computer is the bringer of information and information is seen as "thought" then the computer is "thinking" when it brings us information. Computers and the control of computers becomes a form of power. They symbolize the power of mind, the power to process seemingly unlimited amounts of information with absolute correctness. The basic beliefs of the "information age" and the "knowledge economy" make this power desirable and the belief in this power's necesity.

The whole process leads to a reification of knowledge and thought. Computers enhance our models of the world by supporting the information processing that underlies these increasingly complex models. However, the very complexity of the models hides the fact that they are only models. The precision and speed of the calculations gives a false sense of certainty. In much the same way that rough guess numbers can be mathematically manipulated to multiple decimal place "statistics" the precision of the machine's processing can mask the rough guesses hidden in the programming.

Computer output is totally dependent on input and on the programming. The programming itself, the very thing that gives the impression of a "thinking machine," is determined by a set of pre-selected and precise instructions. This leads to a system's determination of results. The act of organizing for machine processing, the assumption of machine usable categories and procedures, the determining impact of how the software is designed, all shape the output. Moreover, this happens while the machines present the illusion of impartial machine determination, separate from human bias and agendas. This process can lead us adopt a data processing model of "thought."

Information and Ideas

Information is not "thought." Thinking with ideas is not "information." Ideas, historically, originate in other ideas and are validated, or not, by ideas. It's often said that ideas arise from "facts," but in truth, "facts" are constructed by ideas. Ideas tell us were to look for "facts." Ideas separate what "facts" are relevant and what "facts" are not. Culture survives and prospers through ideas, ideas that carry the power to inspire us to shape our world by our actions. Democracy, justice, honor, profit, efficiency and utility are ideas not "facts." In the final analysis it's ideas that create information which in turn supports more ideas and thinking further.

As actors located in society and history, we "think further" the ideas that we have been given. In practice the type of ideas called "generalizations" provide the patterns that connect and give meaning to piles of "facts." This is not a linear, Cartesian process of the orderly accumulation of data. Computers may accumulate data in an orderly process but people "think" through their imaginations. People "think" when their imaginations impose patterns on particular events or facts. These patterns are then subjected to a number of tests, for example consistency with other information, a plausible theory of causation etc. These tests, based on ideas, influence the pattern we see in the facts or events. Our use of these tests makes the perceived pattern grow stronger and more certain or fall apart. This is the process of "thinking" that we use to refine and develop our ideas.

The truth value of generalizations ranges along a continuum from near certainty to wild hunch. Their certainty tends to conform to the amount of data and information that supports the generalization (it also varies with the amount of data or information that does not support the generalization). Informal, usually initial, generalizations are called hunches or intuition. Formal generalizations are called theories or hypothesis. In general, the less supporting data there is or the more ambiguous the supporting data the weaker and more unstable the generalization.

Patterns are in the eye of the beholder. However, these patterns are held together with ideas, ideas that give meaning and explain connections or lack of connections. Ideas are supported or refuted by facts and good ideas have to account for awkward facts as well as those that support them. Facts are also created by ideas. Defining "illegitimacy" creates illegitimacy, passing a curfew law creates the crime and the "fact" of curfew breaking.

Reasoning requires a spectrum of information from more than one source and encompasses more than one viewpoint. This is a major problem for schemes to produce "information machines" because understanding often requires access to non-official views and non-approved information. Since the suppression of alternative views, for either political or commercial reasons, force critics to "re-invent the wheel" the temptation to censor is endemic. Further, the lack of a framework of ideas to support alternate viewpoints can lead to the sidelining of contradictory facts as merely "anomalies" or "irrelevancies."

The great master ideas that shape culture are often the ideas supported by the fewest facts. The great ideas of religion, politics and culture are often so value laden that they can not be substantiated by "facts." These master ideas shape the world by biasing the mind towards established perspectives. They stimulate other ideas and like the foundation of a house structure all that is built upon them. Master ideas such as the divine right of kings or the equality of all men can inspire whole nations and structure whole systems of politics and law.

Ideas, not facts are the integrated and integrating patterns of thought that explain things. They give order to experience (both group and individual). In many ways, ideas are the product of imagination inspired by experience. Information is of course important but it's created and shaped by ideas. Knowledge only comes from thinking and thinking can only be done with ideas.

Ideas take us into other people's minds and link us to their experiences. To understand an idea is to gain an understanding of the lives of the people that created and championed it. The data processing model of thought constrains these abilities down to the limits of what computers are capable of handling. The data-processing model, through the speed and precision of machine calculation, give a false certainty that causes us to mistake a model for reality. The data processing model denies the interplay of experience, memory and ideas that are the basis of all thought. By reducing "thinking" to a machine model, we forget that experience is the raw material that ideas come form.

Using experience as the basis for forming ideas is the process that structures much of our understanding of the world. Science, in the modern sense of the word, is based on experience, the specific and particular "experience" of testing and experimentation, where each experiment gives us a specific and rigorously recorded "experience" of the process and result. History, one of the great systems of knowing, is based primarily on experience and the recording of experience. Despite a shift towards the other social sciences, especially economics, for explanation it's surprising how often we will "explain" something by telling a story of its history.

The alternative to this experience based system of knowledge is a return to the medieval system of knowledge based on authority. Authority that can be publicly proclaimed in the manner of churches or some professional organizations, or authority can be hidden behind a technological mask, tucked safely out of sight in software design or in the choice of criteria for what is "valid."

Our memory or our "history," of our experiences is both varied and fluid. Different groups have had different experiences and interpreted those experiences differently. As events and arguments proceed interpretations of experience are modified and evolve. The shaping of knowledge though a "thinking machine" reduces much of this diversity in experience to the experience and viewpoints of the machine designers and programmers. This is a great deal of power to give to any group. For those who shape our memory and our ideas can shape our lives. Ideas, often produced by the give and take between different groups and individuals, must necessarily suffer if "knowledge" is reduces to data and information, especially, if the selection and organizing of that knowledge is left to those behind the video screen of our "thinking" machines.

Ideas: Different Approaches

Our civilization, in its modern form, has been shaped by how we view ideas and how we control ideas. In the medieval era, ideas were the property of kings and churches. Disagreement was treason and sin. The world had an official shape and an official interpretation. In practice, there was an official place for everything.

That is, until the discoveries of the 16th century brought new and unaccounted for worlds that no church or ancient philosopher had imagined. The great medieval map of reality, already shaken by religious schism, was undermined with each new thing. Intellectually, this stimulus forced society to accept a "New World," a world that had to be seen in new ways in order to be seen at all. By the 17th century, the desperate attempts to explain new peoples by referring to "lost tribes" of Israel or the descendants of "Trojans" had largely failed. New ways of looking at the world and assessing knowledge were being invented.

One new "style" of thinking that came out of this ferment was the questioning style of the Enlightenment. This "style" of thinking was rigorous and targeted. It used close observation and had an experimental spirit. It tried to test assumptions and presuppositions by putting all things to the test of observation and measurement. It's the "style" of thinking we call science.

Science is a form of understanding that uses a process of structured inquiry. However, contrary to what some of the early theorist thought, facts do not speak for themselves. Facts become relevant, indeed only become "facts," in relationship to some system of inquiry. Facts can be "facts" only in relationship to something. The need to categorize, measure and assess prior to investigation adds a "constructed" element to facts. Facts only "speak" through our imagination. It's our ideas that show us where to look for "facts," and how to recognize them when we "find" them.

The data processing model of thinking makes facts superior to ideas. In a sense, it replaces the structuring dogmas of the church or the "philosophers" with a structuring dogma of software and Cartesian assumptions. This approach gains some of its prestige from the success of the scientific empiricists, who approached science by excluding anything that did not lead to "verifiable knowledge" or "certain knowledge." This approach underrated the value of the imagination. It led them to the belief that their systems were "objective" and not subject to the distortions of the system itself.

These approaches lead to the false belief that quantitative research and qualitative research can be separated. In practice most qualitative studies contains quantities even if they are the non-numerical quantities of more/less, faster/slower or increasing/ decreasing. More importantly, these approaches led to the belief that there is such a thing as a purely "quantitative" study. Yet, in practice even a table of numerical results is qualitative in nature. The decision to make the study is a qualitative decision. Our choice of what to measure is also qualitative. The selection of categories and the relationship of categories are qualitative decisions. The interpretation of what the table means is qualitative. Under the neat rows of numbers, there is a necessary and unavoidable structure of qualitative decisions. The data processing model of analysis often masks this qualitative structure by "hiding" it in the software. The qualitative factors then become subliminally locked in the language, the software and the assumptions.

Fantasy of the "Thinking" Computer

Computers have the ability to store information and process it according to strict instructions. Its memory is just storage, like an electronic filing cabinet, its processing is just mathematical applications like an adding machine. This can present the illusion of thinking. The mathematical reasoning (logic) of the machine is necessarily tied to an ideal mathematical model of the world. Circles are perfect circles, data is always crisp with no fuzzy edges, and categories are always exact. In this perfect mathematical world, the mathematical processes of the computer give certainty. (Apples are always apples, an apple = an apple, but in the real world no two apples are exactly alike.) Faith in the machine is akin to the natural science's faith in numbers. Like the spreadsheet cult of finance, the illusion of mathematical certainty means that that the quantifiable becomes supreme while the underlying qualitative assumptions are hidden.

However, intelligence is the ability to use ideas not process data. Intelligence is the ability to creatively use and compare ideas. In fancy, the large amount of data used and the precision of its calculations lend the computer the mystic quality of certainty. In reality, computers only follow instructions and will endlessly reproduce the errors or inadequacy of those instructions. The accuracy of the process is not guaranteed and the computer's claim to accuracy is based on the assumption that the programmers got the process perfect.

This math driven "thinking" has little resemblance to thinking in the real world. The basis of the computer's "thinking" is the formal and constructed world of mathematics, with its built in axioms, unambiguous rules and clear distinctions. This means that the data, the categories and the process must be selectively filtered to eliminate the fuzziness, ambiguity and paradoxes of the real world. This shift moves the computer one step away from the real world and undermines its ability to "see" that reality.

The computer model of "thinking" is related to an old and discredited system of meta-physics. This system is the reductionist system of Descartes' certainty and Newton's universal laws. This simplistic clockwork view of reality matches the computer's actual abilities. The precise series of instructions (algorithms) that govern its every action organize all of its contents the same way that Newton's gravity organized the cosmos. No algorithm = no software = no processing. The data is therefore subordinate to the programming. The programming in turn is subordinate to its own basic assumptions, its pre-set limits and assigned values. As the programmers say GIGO (garbage in = garbage out).

The rigid logic patterns of the computer are effective when applied to situations that have rigid and strictly set rules. This is why they are so good at games like checkers and chess. There are a set number of factors, each piece has strictly controlled abilities and there are a finite number of options. In short, it's the constructed nature of the game's own rules that opens the way for computer "thinking." However, when confronted by choice, preferences and value judgments the machine is stumped. Programmers get around this problem by programming in the qualitative elements for the machine. The machine's "ideas" and qualitative "thinking" are not actually the machines at all. Without ideas and without addressing qualitative factors there is no "thinking" in any meaningful sense of the word. This makes the dream of a "thinking machine" a perpetual fantasy, which no increase in hardware performance or software improvement can deliver.

Quantity Effects Quality

A Tidal Wave of Data

The last fifty years have produced a tidal wave of information. We know more about more things than ever before in history. We are learning more about more things faster than ever before. By the 1960's, this growing mass of information had become completely unmanageable. Fortunately, just as we were about to be overwhelmed computer technology began to address the problem. One has only to look at a 1950's movie to see the way it was. Look at the office in the background! The ranks of clerical workers, often filling huge rooms, each busy doing the same repetitive administrative work. The computer, with its data storage capability and more importantly its ability to retrieve, sort and arrange data gave us just the data management technology that we so desperately needed. Indeed, that was one of the major reasons that billions of dollars were invested in computer technology.

However, the Internet, by giving us "unlimited" access to the world's information, has re-created a problem that the computer had promised to solve. We are again being buried in a tidal wave of information. Research on any subject, no matter how obscure, now brings a flood of web sites and the websites themselves link to an ever greater number of "associated" or "recommended" sites. (In this respect hyperlinks can be regarded as electronic footnotes.) If information and access to information is wealth than we are all stinking rich. If information and access to information is power then we are all very powerful indeed.

However, there seems to be a problem. More information and greater access to information should be making us better educated and better informed. Instead, the opposite seems to be happening. Virtually any public issue or idea of substance is "unavoidably" dumbed down for public consumption. Another example is e-mail, it was supposed to effortlessly connect us and radically extend our span of control. Yet, e-mail has become a burden to many as they face the daily task of wading through an ever-deeper mailbox.

Information seen as a simple flow of raw data is clearly creating problems. The overwhelming quantity of information available is preventing us from sorting out the information we really need from that that is merely available. So great is the flow of data that we are becoming subject to the law of diminishing returns, more data means less understanding. This problem can only be resolved if we make a critical distinction between having great quantities of data and being able to use that data.

It also requires us to reject the myth that information in and of itself is power. Information is useful to power; information helps to establish power. Information and knowledge, at least to a certain extent, are necessary for power. They can be a basis of power. However, neither is "in itself" power. To have power in the practical sense, one must also be able to do as well as know and "doing" requires economic, political and social resources. This is an important understanding, especially in the political sense. Despite many claims about the importance of knowledge, our merely knowing is not enough.

This understanding undermines much of the "information age" hype. Knowing that one does not have a job does not create a job. Knowing that one does not have the money to defend oneself in court does not create the money. Knowing that what the government is doing is unconstitutional does not give one the political influence to stop them. A tidal wave of information by itself will not solve the economic, social or political problems of the day.

Nor will moving data faster, by itself, solve our problems. Indeed, our love affair with speed has shifted from fast cars to fast information. The "instantaneous" speed of e-mails has not noticeably improved their contents, nor has it allowed the harried executive to read or digest them any quicker than snail mail (postal mail). Our preoccupation with fast media – the Internet, TV – is more about urgency than worth. A book, one of the "slowest" and oldest mass medias, is the media that often carries the most important message. In part, this is due to a books greater length. In part this is due to the very time it took to research, analyze and arrange its production.

Despite the fact, that many promises of the new "information age" have proven false; the process is now unstoppable. The irreversible momentum of new computer and communication's technologies is speeding up, not slowing. Computer companies are now trying to introduce a new generation of machines every eighteen months. New software developments are happening even faster. The Internet continues its breathtaking, exponential growth.

However, the notion that technology necessarily equals progress is getting very threadbare. As some of the hype wears off the new "Knowledge Economy" and "Information Age," we begin to understand that technology is a double-edged sword. It improves performance, but also introduces new requirements. It frees us from some tasks, particularly routine bull-work jobs, at the same time if burdens us with the necessity of mastering complex skills.

Information Mobilization

A massive scientific research effort and the new inventions it produced is widely regarded as a major cause of Allied victory in WWII. The war also brought credibility to central planning exercised on a vast scale, for example the "management" of industry and the scientific classification and management of labor during the war. After the war, large-scale management based on vast amounts of information was seen as a positive good. Government, industry and commerce all attempted to continue this wartime progress by implementing policies that used both the latest scientific findings and economies of scale to promote efficiency. Doing this required the ability to find, collect, manage and analyze data on a vast scale. Many of the communications, computer and business machine technologies used today were developed in this era specifically to fill this need.

This post-war expansion of information driven management set the scene for the Internet. The various pieces of technology, the computer, software engineering, hi-speed/hi-capacity cable and a number of other technologies came together in the 1990's to enable the Net. This combination of cheap computer storage and processing power with very cheap telecommunications created the current "information era."
So far, this narrative conforms to a story you have doubtless heard before, quite possibly from a salesperson speaking in breathless awe. Unfortunately, this proliferation of communications ability made mass communications available to "everybody." I say "unfortunately" because not "everybody" had the same

agenda or the same "standards" when it came to information. This lack of gatekeepers has led to a blurring of distinctions between what had been relatively separate types of communications. In short, the good part of the Net, removing the gatekeepers, is also the bad part of the Net; there are no gatekeepers.

This has led to a blurring of boundaries. Political polemics are presented as "scholarship," ads become advertorials and edutainment, and news becomes entertainment and public relations. Without gatekeepers to filter and regulate information, a variety of styles emerges. Without gatekeepers to enforce standards, standards disappear as people use the information commons as they see fit. This passes the role of gatekeeper on to the reader. In a "free market" of information its "buyer beware." The reader must necessarily become more sophisticated in assessing information.

Our tools carry our ideologies imbedded within them, the very forces that led us to develop these tools and not others. The forces that shape how these tools are used are also reflected in the tools themselves. We shape the tools and the tools shape us. One can see this historically in the way that the railroads and the highways broke up the ancient localism of communities or in the way that radio and TV reduced differences in local dialect. The tools of the new "Information Age" are shaping us.

The modern age of TV and the Net has "abolished" time and space. We see and hear about things a world away in hours. We see and hear about the event downtown on the evening news. We "see it"; we have it "reported" by "witnesses," so it must be true. Well - we know better. The modern media filters our information. What we see is what "they" point their cameras at and what the cutting room chose to select. What we hear reported is what someone thought we "should" hear. It is all so very convincing, especially if we hear the same thing from different sources.

Yet, a real world grounding in experience is missing. Instead, we have a second or third hand filtering of experience. This can lead to the construction of a fantasy world that we "all know." The information blitzes of the new mass media's have lent themselves to massive propaganda campaigns. Television has opened a window on a world far larger than we were ever likely to see for ourselves. Yet, it has also opened a world of demagogues and managed news. Information is mobilized for us and against us. Information is denied us or we are paralyzed by a tidal wave of information being dumped on us.

However, the emergence of a new and for the moment "public," mass media has changed the dynamics of the post-WWII information explosion. The Internet's lack of gatekeepers (no shortage of wanna-be gatekeepers) opens new possibilities. More than anything else it makes us players in the information game after decades of being pawns. For the time being there is a new information player in town and it's changing how the game is being played.

Impacts of the New Technology

The new media of the Internet shifts power relationships by making mass communications available to people with few resources. It also shifts power relationships by making things much harder to hide. We hear about things because small groups or individuals go to the time and effort to set up web sites that serve special interest groups or address particular issues. It's not that these groups or interests did not exist before. It's not even that it wasn't possible to collect large amounts of "non-official" information; it was. What is different is that this information can now be shared with a mass audience and kept available for reference by anyone who chooses. Moreover, the costs of doing so are very low.

These are the unintended consequences of technologies intended for top down control. Technologies that were intended to support the "superbrain" computer and allow centralized control of politics and the economy. Instead, we got an uncontrolled mass media. Nor have the expected social impacts occurred. In the 1960's, the experts were predicting that the computer and automation would create a "problem" of leisure, that we would have a four day 32 hour week. Instead, we've increased our work year by 164 hours – almost a full month. Moreover, we are doing it for the same or lower wages.

There are several reasons why this leisure age never happened. Partially, it was because we shipped tens of thousands of hi-wage jobs overseas and allowed foreign manufactures to enter duty free. Partially, it was the demographics of worker surplus as the baby boom worked its way through the economy and mass immigration brought in even more workers to glut the market. Automation and mechanization have also pushed us in the direction of labor surpluses. These factors have kept continual pressure on the job market and reduced the political power of workers. For those using the machines, it has meant more work not less. We have not used the technology to do the same things faster; we have used the technology to do more things. The spreadsheet gets one more polish, the essay gets one more re-write, and the statistical number crunch adds more figures.

TV, once thought of as the new "educational medium," has only partially fulfilled that promise. Cultural, scientific and educational TV does exist, and in quantity, but light entertainment and re-runs dominate. In some ways TV has been both an Agora (a top down Agora were "they" determined what should be discussed) and a coliseum producing contact sports and live wrestling. Public Television exists and often produces excellent and informative programs but its "Donor" system of funding has effectively de-politicized the channel. Animals, ancient history, high culture and such predominate. Few political issues under a quarter century old or currently in debate appear.

Those who saw the possibility of a new republic, or the rebirth of the old dream of a Jeffersonian Republic in the new availability of information, have been encouraged and disappointed. Encouraged because America's extraordinarily narrow political debate is being widened and disenfranchised groups are gaining a public voice, disappointed because this hasn't led to any fundamental system's change. But then, why should it? Information alone is not power, nor is "learning about" the same as thing as "doing something about."

The current gap between public knowledge and public understanding is a serious problem in a Republic that depends on politically active citizens. Moreover, it is impossible to tell if this gap is growing or shrinking. We, as a group, seem to know so much more then previous generations yet we have startling gaps in our knowledge. For example, researchers found that as the Gulf War progressed those who watched more TV actually knew less about the war then those who watched less TV.

(Some have attributed this to a manipulated and controlled press).

Our schools continue to control and filter information as they always have. Partly because students need to have their information controlled and filtered while they learn how to learn, partially because government schools have always been in the business of "producing" the "kinds" of people government wants – or trying to. The current trend towards "the sciences" and "job related" skills is not new; indeed it is at least fifty years old. The "politically correct" agenda is also not new although zero-tolerance and the criminalization of youth are intensifying as the schools attempt at imposing a moral agenda by brute force. (The anti-alcohol moralizing of the turn of the century was probably just as intense and certainly more effective since it helped to produce the disaster of prohibition.)

The new technologies have also had an unexpected impact by never being finished. The machine is never "bought" it just keeps expanding; businesses and schools that assumed their multi-million dollar investments where capital investments are now finding that they are not. The add-ons, the upgrades, the new models, the necessity of ever more purchases just to keep operational have made these investments operating expenses. Information anxiety is part of the sales pitch in this expensive round of eternal upgrades.

This has produced a glut of obsolete technical expertise and a continual shortage of workers trained for the current systems. Large amounts of data become unavailable as the technology for accessing the data becomes obsolete. These technological renewals have become the price of doing business. Those who fail to pay the price soon find themselves left behind.

Information continues to both increase and move faster. Political and social data wars erupt as each side manipulates statistics, facts and empirical opinions. Information which "wants to be free" also "wants to be sold" and "wants to be proprietary." So many "studies" have been done on so many things that there is now a growing field of meta-analysis, the combining of studies for composite analysis.

Problems of too much and too little information, problems determining information quality and problems of learning are all present. The possibilities of the new technologies are here. The problems of the new technologies are here. However, they are both very new and people are still very much exploring how things "might be done" even as the things that can be done change.

Problems of DataGlut

One of the problems of DataGlut (too much information to digest) is that it bogs people down in a mass of information. The amount of information available and the ease with which it can be manipulated can lead to statistical anarchy as various political groups "spin doctor" the numbers. The public is then confronted with a confusing mass of claims and counter-claims of factual "scientific truth." This can create political paralysis by analysis and hinder any effective political action. Under conditions of data overload and contesting claims there is a tendency for decisions to be made based on custom, turf, clout or politically driven opinion poll rather than reason.

One solution to the problem of dataglut is the establishment of information integrity, the confirmation of its accuracy, suitability, timeliness etc. Information must be examined and analysed to determine if it has been skewed by by the researcher's decision about what is measured or how it is measured. This is more than just getting simple facts recorded correctly. (Keeping data accurate and finding inaccuracies is a major concern especially for large databases.) A common measurement problem is determining the suitability of the measure used. For example Gross Domestic Product (GDP), the commonly used yardstick of how well the economy is doing, is merely a measure of activity. Building prisons is recorded exactly the same way as building schools; a bankruptcy sale is no different from any other sale. Consequently, the measure is meaningless unless one assumes that motion is progress and more activity is better regardless of what that activity is. Clearly, basing policy solely on a statistic of this nature is at best problematic.

The problem of data or information integrity is further complicated by the statistics wars generated by special interests. Public relations agencies produce facts for hire, set up phony front groups, hire "grassroots" activists and supply news outlets with pre-written "news." Well and not so well funded think tanks present slanted analysis for public consumption. Indeed the "influence the public" market is a multi billion dollar enterprise. The players with the most connections and the deepest pockets tend to have the most influence. Advertising alone has grown by 2200% per capita between 1930 and 1990.

Quality of information and how to evaluate the quality of information are increasingly critical problems. This problem is complicated by the sheer mass of information coming at people. As they become overwhelmed people stop seeking new sources and try to filter or restrict their current sources. This, very understandable trend, stimulates information providers to try harder to get people's attention and their message across. Extreme statements and positions can be espoused - simply for effect. The freakish and the sensational are highlighted to draw interest and moderate discourse declines. As this happens the flow of information becomes "noisier" and more annoying, often causing further information rejection. Overall, these developments leave us with assurances of information quantity but no assurances of information quality.

Nicheing

One of the consequences of dataglut and information specialization is nicheing. With a web site for everything and an interest group for every issue there is a tendency for people to narrow their inquires to their own interests and exclude everything else. To a certain extent, this is perfectly reasonable and effective. After all, responding to complexity and information overload with specialization is a traditional method and gave us the various trades and professions. However, beyond a certain point specialization can lead to fragmentation and polarization. Communities of interest that become disconnected from mainstream issues and problems tend to underestimate how dependent they are on the shared elements in their society. The common good may be something of a cliché but it also has an element of truth. No group is independent of the society in which it has its existence.

The issues of mass culture and the benefits of mass culture itself are debatable, particularly when it is difficult to tell were mass culture is going. We know historically that mass culture is recent, arguably less then a hundred years old. It was the railway and latter the highway that first created the mobility that helped break down cultural isolation. The impacts of mass communication, first newspapers, then radio and TV connected people in a way never before realized. Most people forget how the advent of radio in the 1930's totally transformed the cultural isolation of farm and small town America. In communications terms we have only had a true mass culture since the end of the Second World War.

Mass culture, stimulated by the shared events of WWII, probably reached its height in the 1950's. However, the narrow conformity, rigidity and stuffiness of the era generated a massive, and spontaneous, rejection of "mainstream" culture in the 60's. Moreover, despite a return to more conservative attitudes we have never truly gone back to the conformity of the 1950's. In part, this has been based on increased communications and the inability of the elites to turn back the clock. In the 1990's, the mass media itself is changing. Mass TV still exists, although it is rapidly breaking into niche markets like MTV or the Sports channel. The magazine market is dominated by thousands of special interest magazines and the great mass-market magazines like Saturday Evening Post are a thing of the past. Only the great newspapers remain as a mainstream, mass-market media.

The Internet has without doubt contributed to this nicheing trend. The different interests were, of course already there, but the Net has enabled their rapid diversification. This trend has been accelerated by the use of technology to identify and address niche markets. The use of computers to find patterns amide a mass of data (data mining) has enabled advertisers and interest groups to target their audience. This ability has supported the proliferation of identity politics and special interests. It has also raised the possibilities of broad alliances. However, the non-elite, non-mainstream groups have largely ignored the possibility of alliances in favor of a quest for purity. After all, why join with people you partially agree with when you can concentrate on those that exactly agree with you. The result has been political fragmentation and infighting. It is unclear, if anti-establishment groups will ever achieve the influence of big business and elite coalitions.

In 1964, President Johnston declared a War on Poverty as part of his Great Society Program. A central part of this effort was the first comprehensive social mapping of the US. Twenty-nine agencies combined over a billion pieces of information to reduce the country to blocks of 1500 households arranged into forty categories of neighborhood type. This is now a routine approach to marketing and politics. However, this geo-demographics approach is ultimately based on stereotypes. It's also problematic because individuals "belong" to so many different political, social and demographic groups. This raises the problem of which grouping should be selected for labeling purposes.

Nichification can undermine common culture. This is especially true if there is little cross communication and can lead to a scarcity of common viewpoints. Without some common ground of analysis and interest, the overall good can be lost in heated special interest fighting. Nicheing can mean a turning inward to a narrow focus while Dataglut inspires a tuning out of everything else. The Net actually encourages this process by providing "smart agents" within web browsers that filter out selected content. The Net tends to support previously disenfranchised groups, while it subtly undercuts larger coalitions.

By shifting us from a mass culture to a niche culture, our technology is working against a shared political economy. Advanced communications promote the lobbying of opinion poll following politicians. However, this very democratic process is being exploited by astro-turfing ("grass roots" organizations that are organized fronts for special interests). Some companies are also ordering their employees to send in mail and e-mail supporting the company's political agendas. These practices are sometimes accompanied by disciplinary monitoring of the employees. This sort of activity can have a significant impact especially if leaders are leading from behind. Sensitivity to polling results based on "opinions" can also yield the political initiative to those who control the media that forms much of those opinions.

Cultural Impacts of the Information Economy

The "new economy" has decreased leisure, shifted the workforce towards temporary jobs and decreased job security. All of these changes have tended to lower political involvement as people shifted their priorities to simply getting through the day and paying the rent. Despite the promise of a new Jeffersonian Democracy based on free speech and public speech, we have seen a decrease in political participation. This is not terribly surprising when we remember that the basis of Jefferson's idea was economics not information. As originally proposed each person would receive 50 acres of land creating an economically independent, politically active yeomanry. This independent yeomanry would form the political foundation of the Republic. (America in Jefferson's time was an agricultural nation and land ownership equaled economic security and independence. We might also note that the modern corporation with "super citizen" powers and rights did not exist until well into the 19th century.)

Modern communications, by linking our computers into one system, has opened the doors to information piracy, computer viruses, destructive hacking and identity theft (stealing a person's identity information and then committing fraud with it). Data-valence, the surveillance of people or groups by monitoring their data is also a growing problem. The data mining of large amounts of information is also being used by various groups to target select audiences for propaganda and commercial purposes. Besides making the collection and distillation of consumer and demographic data cheap and easy, technology has made outright surveillance itself very affordable. These trends have raised concerns over the use of special interest politics to disrupt the larger political process. Concerns for privacy and exploitation are also increasing at the very time when the political influence of individuals is declining.

The 1974 Federal Privacy Act severely restricted government data collection. However, it exempted business from any restrictions making information gathering and sale a lucrative industry. For example some companies are now collecting school records in bulk and selling the information to prospective employers. This approach to privacy in the commercial sector is also creating problems with the European community who have enacted laws protecting individuals from unwanted commercial surveillance.

Another pressing concern is the faith that some groups have in statistical data. Some people actually believe that "profiling" is reality, that we can really assess potential criminality or political attitudes by measuring secondary data. This is on an intellectual par with determining who the "nice girls" are by measuring skirt length. Statistical profiles, even when accurate, are only "true" in the aggregate sense. They don't tell us about any particular individual with any reliable degree of accuracy. This means that the current trend towards "profiling" (a politically correct term for stereotyping) is a return to the old practice of criminalizing according to race, ethnicity or other group membership. This makes the current and growing practice of identifying people by secondary data highly questionable (for example patterns of bank transactions, what books they own or whether they have "gang clothing" – whatever that might mean?). The growing integration of official computer networks makes this a problem that is spreading across government.

The flood of information can effect out ability to perceive real events. Reasoning and careful analysis takes time and a certain amount of dispassion. People who are busy and overloaded with information are prone to be distracted by emotion or opinion. This can lead to very superficial analysis or analysis that simply rejects any awkward bit of information. Under these conditions, lies and dis-information can push out truth. Indeed, we have been subjected to decades of public policy by lie. The decision to go into Vietnam was based on an incident in the Tonkin Gulf, which we now know never happened. The final impetus to the Gulf War was a fabricated story of murdered babies and stolen incubators. Welfare was dramatically reduced with the propagation of the "welfare queen in her Cadillac" myth. (Pres. Reagan told the story in the campaigns of 1976 and 1980. The woman was supposed to have 80 aliases, 30 addresses, 12 social security cards, and gained $150,000 by fraud. This was sold as a "typical" case justifying massive cutbacks. No such woman ever existed.)

This power of anecdote, the compelling story, is the tool of demagogues. It has emotional grip, it clarifies the issue and it can often mislead as an isolated incident is characterized as the "norm." Such stories now abound in our political debates.

(We might pause and note here that these two problems are of importance to researchers. An anecdote without statistics prevents the researcher from determining if the story is "typical" or one in a million, an important distinction for public policy to make. The use of statistics to create a "story," the stereotype of someone who is "typical" of those statistics can also be very misleading. Indeed, it invents reality out of numbers abstracted from reality.)

We also suffer from deliberate deception. Infomercials present advertising as journalism. Advertorials present advertising as editorial content. The short sound bite style of much of today's mass media makes it increasingly hard to tell what exactly it is that we are seeing.

For the modern researcher, dishonesty, agenda scholarship and dis-information have become major obstacles to good research. The Internet opens the possibility that people can by-pass the top-down filtered information that increasingly dominates the other mass media. However, this possibility is not simple, effortless or guaranteed. Formally, the possibility is there and anyone can take advantage of it. Effectively, only those who understand the issues and have the intellectual skills to sort things out can make use of the opportunity.

The Net may favor Libertarian ideas like decentralization and free markets but it can also undermine common discourse. This makes the Net a problematic political tool for changing the world. However, the Internet credo that this is a turning point in history with tremendous economic and cultural impacts is at least partially true. The arrival and growth of a new mass media available to "every-one" is a significant development. However, it's not just a tool for Libertarians or Jeffersonians. The Net is also a tool for those who currently control the other mass medias. Indeed the existence of an unregulated mass media is being used as an argument for de-regulating everything. The cheerful anarchy of the Net is presented as a model for a deregulated anarchy right across the policy spectrum. A curious argument, when the Internet's development has been so dependent on government initiative, funding and regulation.

The age of information is undoubtedly upon us. It connects us to the world of information as never before yet at the same time disconnects us from the world of experience. This has led some to confuse information with knowledge. More then anything else, it has led to the confusing of quality with quantity.

Censorship

General Nature and Purpose

Censorship is not about morality, nor is it about crime. Censorship is about power. Despite the rhetoric of how awful, dangerous, or worthless something is. Despite rhetoric about how something disgusts them or frightens them, no one asks the government to impose censorship on themselves. If they personally did not wish to read, hear, or see something they could simply put the book down or turn the device off. Censorship is always about not letting someone else see or know something. Only those with a claim of superiority and a desire for power over other people institute censorship. It is "youth," "the lower classes," or "those people" who are "not ready" or "too easily excited" or "unable to handle" the material involved. Censorship is an act of power with the deliberate purpose of restricting less powerful people's knowledge.

Censorship is normally defined as restrictions on knowledge or speech. In practice, the line between censorship and prohibited activities is very fuzzy. Structurally censorship is very similar to laws against consensual activities or "victimless" crimes. Further, censorship usually overlaps into the area of proscribed activities. For example, laws against pornography must necessarily affect sexual activity associated with pornography. Though often separated in academics, the distinction is blurred because so many of the laws against "consensual crimes" have strong elements of censorship in their justification and enforcement.

Censorship is an act of religious, cultural or ultimately, political power. It's driven by religious, cultural, economic and political motives. Someone, or more often, some group of people have decided to dis-empower some other group by denying them information or the use of public discourse. By doing this the censored group is also denied the freedom to defend its position. This is usually justified with arguments that it "protects" the individuals involved or that it "protects" society. In practice, it protects the privilege, position and agendas of the censors through the enforcement of ignorance. This amounts to the imposition of political agendas by intellectual fiat. (The old law that made it a "crime" to pass information about birth control is a good example of social agendas – increasing population, punishing "bad" girls etc. - pursued through enforced ignorance.)

However, these agendas create problems for law enforcement. The first problem is the highly subjective nature of the material. On person's pornography, is another person's erotica, is another person's picture of girls on the beach. The problem of determining where the actual lines of "criminality" are makes censorship enforcement highly arbitrary. The subjective nature of censorship is often reflected in the vague or sweeping way in which the laws are written. For example, the current "child" pornography law applies up to age eighteen. Consequently, we have a law that categorizes young adults over the age of consent as "children." (We can note here that restrictions are easiest to apply to people who have diminished legal rights. People who can't vote (or do not) and people who can't make big political contributions are an easy target for "morality" laws. Such laws also "condition" youth to accept these restrictions as normal.)

Our law enforcement system is based on a model of crime victims reporting the crime. Crimes are supposed to have a victim who will report the crime to authorities. The authorities, having been notified of the crime, then respond to the citizen's complaint. However, when the "crime" is victimless or consensual there is no victim and people who want the censored material do not report their having it to the police. This means that the authorities have no complaint and have to find the "crime" in the first place. Unfortunately, the people engaged in these "crimes" do not want to be caught and punished.

This sets up a continuing round of move and countermove as the authorities try to catch people and people try not to be caught. As people conceal their activities by moving them out of the public sphere the authorities find their enforcement thwarted. The only way the authorities can then continue the game is by making greater intrusions into the sphere of private life. The idea that this intrusiveness is only directed against "criminals" is often stated but not true. The "unreported" nature of the offense makes everyone suspect. Since the "crime" is not reported directly, the authorities must become far more intrusive and begin seeking secondary evidence as an indicator of "crime." This is particularly true for "thought crimes" or "information crimes" like censorship.

This pattern has been very prevalent in the "War on Drugs." (The "War on Drugs" has often resorted to censorship, dis-information and other efforts to indoctrinate the public.) Civil liberty after civil liberty has been sacrificed to "win" the war. Immense amounts of secondary records, such as bank records, are examined to try to find "evidence." (Bank records used to be private unless there was a search warrant; now new laws are causing banks to simply turning over all records to the government as a matter of routine.) Doctors are now required to report patients and schools routinely engage in mass testing. A massive system of paid informers (occasionally agent provocateurs) is now in place, all to detect "evidence." This trend continues, with proposals for new laws making it a "crime" to present information about manufacturing drugs.

This trend towards ever greater intrusion is almost always justified as a "necessity" and it's usually implied that it's only "for the crisis" or until "the war is won." Indeed, we've had three decades of being one civil liberty away from victory. In each case, the sacrifice of a civil liberty or a due process has proven massively ineffective. It has succeeded in producing a host of "new crimes" and a host of prisoners convicted of these crimes. However, it has not solved the original problem.

These law enforcement difficulties with consensual crimes and censorship are complicated by the very number of "suspects." In practice, this amounts to the entire population that must be "watched" and "disciplined." (For authoritarians this requirement is not a problem, it's a good thing.) Since there is no direct measurement of the activity (no reporting), secondary means, often through mass surveillance, are used and they tend to be crude. The maintenance of huge nationwide databases to track irregularities that "might" indicate a crime sounds wonderfully easy and powerful, but it's immensely costly. Keeping the records of secondary activities accurate and up to date enough for legal purposes is an administrative nightmare. These problems then lead to further expenses and difficulties. These schemes to track compliance with secondary data also tend to falsely incriminate some while those actually involved avoid attracting attention.

This surveillance problem is so difficult and expensive to deal with that censors usually try to enlist searchers, voluntary or otherwise. This often leads to mandatory reporting schemes by completely untrained amateurs. For example under the child pornography laws parents are now being arrested for taking a picture of their kid in the bath or at the beach. In practice, the low wage employees who process film have replaced the Supreme Court in the determination of "what is pornography?"

Censorship as law enforcement is clearly problematic. However, from the viewpoint of power it's very effective. Censorship deprives the people involved of any public defense of their viewpoint. This leaves the field open for the censors to promote any version of reality they desire. The "Reefer Madness" approach to marijuana is typical of this (Reefer Madness was an early film depicting marijuana as highly additive and unleashing murderous impulses – neither of which is true.) This prohibition on speech helps to promote ignorance and intolerance thereby helping to keep the censorship in place. (The UN is currently promoting a law making it a crime to say anything positive about illegal drugs.)

The "need" to catch censorship violators provides an excuse for further restrictions of liberty and due process. Each new "drug exception to the Bill of Rights" or "child exception" to free speech reduces the freedom and liberty of the general population. At the same time, it re-enforces the top down nature of power that created the censorship in the first place. This makes the censoring group more powerful. As power shifts it becomes easier to impose more top down rules and create more "exceptions" to people's rights. The collapse of British civil rights and due process from exemplary at the turn of the century to well below European standards today is an example of this process.

The top down nature of censorship usually guarantees that the preferences of the politically active and powerful prevail. These powerful elites use these restrictions to gain greater control of society at the expense of weaker groups. The rhetoric of being "just one more civil liberty" away from victory is demonstrably false – though popular. The massive reductions in civil liberties and due process justified by the decades old "War on Drugs" has had little if any effect on drug use. Its bodyguard of censorship laws, dis-information and "official attitudes" has also failed to solve the problem. Indeed, the ignorance created by censorship and "official truth" has tended to cripple any rational discussion of drugs or drug policies. These simple punishment approaches have been unable to control drug use, even in the toughest maximum-security prisons in the country. This makes the "War on Drugs" an excellent example of how censorship and prohibition tend to reinforce each other.

Censorship laws also create a climate of fear, bigotry, oppression and conformity. The demonizing of some activities or groups promotes the belief in demons. (This is not to say that the activities censored do not create harm, they sometimes do, but harm is specific and likely to be reported. General proscriptions that create "victimless crimes" cannot rely on voluntary reporting.) The ongoing climate of fear and suspicion pressures people into conformity, at least in outward appearances. This amounts to control by intimidation, a very restrictive and arbitrary way to run society.

Even when the information censored is potentially harmful; putting the information to use is often illegal. (It is hard to think of information that is, in itself, harmful – just by the knowing.) For example bomb making - much less using is a crime. Censoring bomb making information is supposed to make these crimes more difficult by denying information to potential bomb makers. This implies that it's easy to prohibit such information – just pass a law! However, censoring a few books or web sites will not deter any but the most casual. To prevent a serious person from getting the information (and someone who is intent on breaking a law is usually serious) would involve the censorship of all books and libraries. It would also entail the suppression of virtually every chemistry textbook in the country. It's easy to see that the cost of this project, if undertaken seriously, would run to millions of dollars and still have little prospect of success. It would also have massive co-lateral costs, for example the inability to teach anybody chemistry for lack of textbooks. This gap between the "simple censorship solution" and the massive complexity and cost of actually doing it is very common. In part, this approach stems from the "rule following," "law abiding" nature of most middle class lawmakers. Rules and laws restrict them – therefore they must "restrict" everyone – therefore the problem is "solved!"

The fact that much of censorship is based on religious belief makes the actual site of censorship a subjective decision. In practice, this decision is made by the powerful, who imposing their value judgments through legislation. The argument of societal protection is also problematic. Most information is simply not that dangerous to society (dangerous to elite agendas is another matter). Further, in practice we must consider whether the cost of censoring certain information is greater than the potential damage. By this, I do not mean the cost of enforcement; I mean the cost incurred because the information is not available and the cost of lost options. The government crackdown on "homosexual pornography" was instrumental in firmly establishing the Aids epidemic. Deprived of any way of spreading explicit warnings the gay community was deprived of the medical information it needed to prevent the spread of Aids.

The whole issue of censorship assumes a hierarchy of merit, which can only be instituted in a hierarchy of power. It's an expression of elite distrust of the less powerful. It is without exception instituted in accordance with elite agendas and in support of elite power, political or other wise.

Shaping Society

Censorship's purpose is to shape society by controlling information and ideas. It relies on the fact that most people simply conform to the society of their upbringing and its effectiveness relies heavily on socialization. It's our upbringing and the ideas that we carry that make us a Kansas City real estate agent or an Italian shepherd. There is nothing wrong with this; it's the normal process of socialization. It's also strongly affected by our ability to learn; for example an Italian shepherd who learns to be a Kansas City real estate agent. Censorship is an attempt to push this very natural process in the direction of social engineering. The young are denied certain information and ideas at the same time that there is an insistence that they learn and conform to other ideas. This "thought control" can extend to the use of the young themselves to monitor and report on their parents. Parents often refrain from telling their children certain facts or viewpoints because they are afraid that the children will get in trouble or get them in trouble. There is considerable evidence that the US public schools are moving in the direction of becoming "though police." This is particularly noticeable in the use of "student information questionnaires" that demand information on parents and in various "zero tolerance" policies that apply to opinions as well as acts.

Society in the "natural sense" of human societies tends to be very resilient. Members of a society who seriously flout its dictates find themselves excluded and penalized. This is considerably different from some group deciding that they are the "guardians" or "arbitrators" of society's standards and then using political and legal muscle to impose their will on others. We might also note at this point that this particular approach assumes that there is "one" society. In the cultural sense, this is not even close to true. Whatever the preferences of political rhetoric, the country in practice is a kaleidoscope of varied and shifting "societies" each with its own "truths" and customs. Censorship laws that deal in subjective judgments about what people should know are an appeal to centralized power. They constitute an elite consensus on how society "should be" once the elite has "shaped" it.

Censorship laws are an imposition of government morality. It's an attempt to prevent us from finding out by ourselves. Its objective is to make our thinking conform to the intellectual hegemony established by the censors. Censorship is often accompanied by media accounts that sensationalize without informing. The official efforts to promote revulsion and fear are designed to prevent any desire to investigate and learn. By proscription, ignorance becomes the political base for policy.

Censoring the Net

The leading edge of censorship on the Net has been pornography. This is not surprising. Playing the sex card is the easiest way to put the issues on an emotional bases. It appeals to a large number of people who are "concerned about sex." The censorship sponsors can also claim that they only want to "protect the children." Since children obviously need to be protected (when and from what is not so obvious), this "protect the children" rhetoric is hard to answer. Who wants to argue, or can successfully argue, that we should not "protect" children? (We can note how quickly and easily this debate slips into a format of all or nothing. In practice, there is a range of opinions on when, how, by whom and from what children should be protected. We might also note that the government which is so "terribly concerned" for children's welfare has an appalling record of care for the children who are under its care and completely under its control (about 500,000 nationally)). However, attempts to censor "obviously improper" sexual material are not restricted to censoring pornography. These attempts, by their very nature, would establish the right to impose censorship and would create an authorized body of censors.

Most proposals for Net censorship deal with material that is available off the Net and uncensored. They are justified by the claim that the Net is different because it's "in the home" and therefore requires special handling. This sounds good but does not stand up to much scrutiny. Unsupervised "children" (carefully undefined – does this mean young adult?) have little difficulty getting these materials outside the home. Actual "children" are more easily supervised. Practically, one would think that parents would have a greater ability to supervise the Net, which is in the home than to supervise material that is readily available outside the home.

In reality, it's a highly emotional wedge issue. Something designed to establish the necessary precedents. We have seen the largely unsuccessful efforts at "pornography" censorship followed by attempts to censor "hate speech," "holocaust revisionism," "drug information" and "terrorist" information. (I put all these terms in brackets because I don't know what they mean and neither do you. We may think we know, but in fact the terms are defined so loosely (when defined at all) and used in so many different ways that, we really don't know what is meant by them.) For example, what is one to think of "hate laws" that apply to whites hating blacks but not the reverse? Even if we agree that these laws are necessary, the way that they are written makes it clear that there is an underlying political agenda. Under these laws only some "hate" is really "hate," other "hate" is "not really hate." If it's not OK to "hate" homosexuals, is it OK to "hate" guns?

There are, of course, a number of legitimate positions that one could take regarding the "need" for these laws. However, they are clearly using terms like "hate" in such a politically colored way that the terms become little more than code for political agendas. They hide as much as they reveal. This can make for very problematic laws, especially as time brings in different administrations and changing agendas.

An example of the Internet's impact on efforts to impose censorship was the Internet community's reaction to the 1995 pornography hoax at Time magazine. This exclusive cover story created a sensation by exposing how the Net was completely dominated by sexual material. The time's article stated that 83.5% of pictures on "Usenet newsgroups" were pornographic. The report also credited the prestigious Carnegie Mellon University for the study. Other major media outlets quickly picked this up, with calls for censorship and criticisms of free speech. However, researchers familiar with the Net began to congregate electronically at the Electronic Frontiers Foundation (EFF). They quickly found (and made public) that the source of the "University study" was a paper by an undergrad. They also quickly established that less than one half of one percent of the Internet dealt with sex. Faced with these revelations, Time Magazine was forced to make a retraction of its story. (The full story of this hoax is still available on-line at the EFF.)

However, once the principle of Net censorship is established it can be expanded to include anything offensive to the "authorities." This can be done directly through censorship laws or it can be done indirectly through "voluntary" standards. This process can also be pursued through secondary means. The courts are increasingly being used to "enact legislation" by judicial fiat. For example the current lawsuits against tobacco and gun manufacturers are designed to impose policy by imposing legal costs and penalties that force behavior. The government has already used this method to bypass the legislature and in the case of Smith and Wesson (a gun manufacturer) has used the threat of lawsuits to force a "voluntary agreement" concerning business practices.

The government has also taken the astonishing step of passing regulatory laws that allow enforcement by private lawsuit (for example the Americans with Disabilities Act). This process allows special interests to use the law in any way they want. This can put people in court against the government due to a lawsuit issued by a private (and interested) party. This approach to regulation is extremely dangerous because it bypasses responsible government (the legislature) and places enforcement of the law in private hands. Given the government's inability to get Net censorship into legislation it's entirely possible that these methods will be used as alternative approaches to censorship.

Private companies can also censor the Net. This can be done indirectly by raising the rates for services (for example political chat sites or BBS), and thereby crowding out less "commercial" sites that can't afford the rates. This, of course, is not "censorship" it's just business and it's "just a co-incidence" that it tilts the Net in the favor of those interests with money. In essence, Internet access providers can freeze out non-commercial sites by charging rates that only a commercial operation can afford. The crowding out of non-commercial sites is a pattern of information control that has already happened in TV and radio. The government's abandonment of ownership and control of the Net to private interests enhances this possibility.

Censorship can also be instituted by the owners of Internet access companies or by the owners of the Net computers and cable networks. The government's current policy of completely privatizing the hardware infrastructure of the Net opens the Net to commercial control of content through company policy. For example, AOL the largest Internet access provider (22 million customers) treats guns the same way it treats sexually explicit sites. They simply remove any that they find offensive. As far as the company is concerned there is no difference between "pornography" and second Amendment rights (right to keep and bear arms). Both are "too controversial." We can note here that while American citizens have rights in relationship to government they have very few rights in relationship to employers or private enterprise.

Can the Net be Censored?

It has become a truism that one cannot censor the Internet. Internet Libertarians usually justify this belief by defining censorship as an obstruction, then asserting that the Net has the ability to route around any obstruction. This is true of obstructions, but censorship is not an obstruction or "damage," its management. In truth, governments and corporations are already censoring the Net. Governments as diverse as France, Saudi Arabia and Australia have imposed and enforcing content restrictions. There are a growing number of access providers who impose content "standards" – code for the censorship of anything they find objectionable.

Although the Net is a web of routes by which information can be sent and accessed; it also has some "chock points" that information must go through. The most obvious is the Internet Service Providers (ISP's) themselves. These commercial entities are the access point were users log onto the Net. Regulating the companies that provide them is an easy way to control these access points. These companies are far fewer in number than the millions of Internet users. They also have a large equipment investment to protect; making them vulnerable to government pressure and politically motivated lawsuits. There have also been a number of successful prosecutions by government of ISP's because of messages sent through them or content on them. It's significant that congress has been very reluctant to provide common carrier protection to Internet providers. (Common carrier protection provides protection from prosecution. For example, telephone companies are not legally or criminally responsible for message sent on the telephone lines.)

The physical infrastructure of the Net also offers sites for control. The Net may exist in a mythical Cyberspace but it operates on very real computers and cable networks. Control these, either through government regulation or through commercial policy and you effectively control Net content. The government, following the precedence it set with radio and TV, is opting out of publicly owned Internet infrastructure. Since American's Constitutional Rights only apply to government this move puts the Net (a public resource) under private control with no guarantee that free speech or other rights will be respected. Private ownership of the infrastructure establishes effective control of the Internet in private hands. These private interests are very quickly re-directed development in directions that they find financially and politically favorable.
There are also central elements of the Net that private or government interests can use for censorship. One of these is the Internet's Domain Name System (DNS). This system controls the addressing of Internet messages. Since there is, only one system of addresses

the DNS system must necessarily be one of centralized control. Each machine – yes every single computer on the Net including yours – must have a separate, accurate address. Without an address your computer can neither send nor receive information. (Stated numerically e.g. 234.34.45.56 an address is called an Internet Protocol address (IP) the same address in a form that humans can remember would be "roger @ bogushome.com" this form is called the Domain Address. For the user these amount to the same thing.)

To use the Net you must find the site you are looking for and it in turn must find you to respond. This process begins when your machine contacts a computer called a Domain Name Server (DNS) which is usually located at your Internet Service Provider (ISP). This machine will contact a computer called a Root Server to find the address you are looking for. The Root Server in turn will contact a domain server (e.g. .com or .edu) which will find the actual address within that domain. This address attaches to your message when the computer sends the message. (Root servers and domain servers are central nodes in the net and can be sites for censorship.)

We can see that, unlike the Net itself, this addressing procedure works on a hierarchy from "high Level" domain servers like .edu to your local Internet Service Provider. The US government used to operate these root servers in conjunction with a voluntary group called the Internet Assigned Numbering Authority (IANA). This body managed and co-ordinate the addressing system. In 1992, the US government hired a private firm Network Solutions, Inc (NSI) to take over management. However, as the value of Internet addresses rose and problems developed over ownership of particularly attractive site names, the government decided to privatize the addressing system.

This led to the creation of a private non-profit corporation (Internet Corporation for Assigned Names and Numbers, ICANN) to take over all address management on a global scale. This organization has taken over the management of the Internet Root Server that controls addresses and is tasked with introducing "competitive, market mechanisms" for allocating Internet names and addresses. Proponents are calling this re-organization a "purely technical" exercise to upgrade the management of Internet addresses. However, this entity is likely to come under enormous pressure to serve political aims. As the organization that controls and administers addresses, it has literally the power of life and death over any Internet site. Since this power also extends to the domain servers themselves, ICANN can regulate access to the servers that give out addresses. For example, a desire to create an economic embargo could result in all of a country's commercial sites being isolated from the net. Or one could specify that all commercial sites must go through one root server which must conform to any regulations or restrictions that ICANN imposed.

This is an immense source of power. Any sites that ICANN does not list would be effectively isolated from the Net. This is a very tempting situation for those who would control the Net for social, commercial or political purposes. This control is also international in scope, which raises questions about who will regulate ICANN. ICANN is already imposing registration fees on each new address. It's also planning to impose reporting requirements on domain name holders and cancel anonymous servers. This is the exact opposite of how the Net was created and in fact establishes worldwide Internet governance. It's also unclear who will set policy or by what process policies will be determined.

Clearly, the forces that see the Internet as a threat to their interests are not happy. For almost three decades, they have imposed uniformity on public thought and successfully kept America's political discourse within extremely narrow bounds. They are now dealing with a dangerous outbreak of "excessive democracy," especially in the area of free speech. The growing number and sophistication of political sites criticizing government policies and corporate privilege is a trend, which they do not approve. Moreover, after thirty years of controlling public discourse they have no plans to let this unsupervised activity continue. The Net and all its potential for opening up political discourse is being threatened by groups that want dissidents silenced, critics deprived of an audience, social agendas unchallenged and conformity of "proper opinion" restored. Politically America is facing the question "Now that the Cold War is over – can we talk?" For some the answer is no.

Or as Thomas Jefferson said:

"I know no safe depository of the ultimate powers of the society but the people themselves; and if we think them not enlightened enough to exercise their control with a wholesome discretion, the remedy is not to take it from them, but to inform their discretion."

Research

Good research is supported by an understanding of how the research process "works" and by an understanding of how research is structured. It's possible of course, to simply look around and try to "find stuff," but researchers are far more effective if they understand the structure of what they are doing. There are also places were the researcher can use rules of thumb or "cook book" approaches, but serious research needs grounding in well-structured theoretical foundations. In this chapter, we will examine how research is put together and the importance of the different parts. (There are numerous books on the mechanics and methodology of research; this book deals with the ideas that structure research.)

Theory

All research is theory based. This is not necessarily formal or even explicitly stated theory, it can be very informal or even largely unexamined (I do not recommend using theories you have not examined!). What is a theory? Despite much mis-teaching in schools a theory is simply a story that explains what is or should be how things came to be or how they work. Without a theory, even if it is only a tentative theory, there can be no research. Without a theory (a story to explain), the researcher would have no idea where to look, or what to look for, or when to look, or how to recognize something important when one found it. Researchers sometimes call a preliminary theory a hypothesis.

Yet theory is not the first step in research, the first step is human will. Someone must desire the research done. There is a reason they want that and not some other research done and often there are preferences as to how they want it to turn out. Despite much talk of "neutral" science or the "disinterested" observer, this human motivation at the heart of research is not a bad thing. It's just how people operate in the real world of human interests. Recognizing this is important. Recognizing the basic fact, that there are intents and desires in all research helps us to sort out the author's agendas.

Once the researcher has identified the purpose of the research, they can begin their efforts with an initial theory. This can be anything from a well-established theory to a hunch. For most investigations, there is usually a body of theory to guide the research. Economist do research using economic theory, criminologist use criminological theory etc. Of course, it's not quite that simple, there are qualitative and subjective factors in the choice of the appropriate theory. The researcher might also develop their own theory based on their experience and knowledge in the subject area. This process sounds very orderly and logical, but hunches, intuition and guessing play a big part in determining the initial direction of research. In some cases, the initial theory is simply a leap of creative insight. In all cases this initial theory will give the researcher guidance about were to look, what to look for and how to recognize it when they see it. It's very common for this initial theory to be at least partially wrong.

(Author's note: In my personal experience as a researcher and analyst I find that I go through a cycle of theory building and collapse. What happens is that I begin with one story (theory) to explain things and will go on until my investigation produces evidence that makes that theory unworkable. I will then build a new theory that includes the new insight into the problem and begin again. This is an informal process of "learning about" as research progresses. It's common for a researcher to go through three or four build and collapse cycles during the research project. This is not failure; this is simply the normal process of research and analysis. For example: I once did research on immigration to Canada and the US. After a fair bit of initial study I discovered that all the material I was using assumed that the US and Canadian immigration practices were the same. As it turned out this was not true in several critical areas. I had to rethink the whole approach and analysis of the paper to accommodate this new and critical information.)

Every theory is made up of a number of parts. Again, theories need not be this formal, but the researcher should be aware that these parts are there. To use a theory intelligently, especially a complex theory built up by a discipline, it's important to study how it is put together and be able to identify its elements.

The ultimate basis of our theories is Metaphysics, our basic beliefs about the nature of the world, life and ourselves. Unless one is doing research in philosophy or religion one is unlikely to delve very deeply into metaphysics. Questions like "what is life?" and "what is reality?" are in the area of metaphysics. In practice, an awareness of the human nature of human "knowing" and the imponderables of the human condition helps to keep one grounded.

Ontology

Of more immediate concern to the researcher, assuming one is not a meta-physician or a mystic is ontology. Formally, ontology is a branch of metaphysics that deals with existence. For our purposes, the ontology of a theory means what is and what is not included in that theory. Ontology is part of every theory. For example, some theories recognize direct divine intervention as a possible cause for events. Whether a theory includes the possibility of direct divine intervention or not is part of its ontology. This is also true for other entities such as the "nation state," "invisible hands" or "evolutionary forces."

For example, the "nation state" can be seen as a political construct that exists only because people "will it"; it has no existence "in itself;" similarly the other "entities, such as Adam Smith's "invisible hands" can be challenged as to whether they actually exist. Technically, the confusing of an abstract concept like "nation state" with concrete reality is reification, as in "the US desires peace." The people or government of the US may desire peace but the US is a political abstraction used to designate a certain area of governance, it does not "desire" anything. To use a theory that included what the "US wants" would require an ontology that included a conscious, self-aware entity called the United States. If we reject a hypothetical "US wants" entity we are led to the more accurate question of which parties, people or groups want. This is a much more discerning analysis of what is going on; tied directly to what we will or will not accept in the theory's ontology.

Researchers need to be aware of the ontology of a theory. The difference between many theories and viewpoints hinges on their ontology. In practice, ontological elements crop up in arguments, often as a sort of short hand for a complex set of ideas. For example, "market forces", "killer instinct", "manifest destiny" or "technological imperative." When used as causation or descriptive elements in a theory they can hide as much as they reveal. Researchers need to examine carefully what the author is referring to and how they are using these "entities." Another way of looking at ontology is to ask "who are the actors in this play and what are their parts?"

Epistemology

Another basic element in theory is epistemology. Formally, epistemology is the branch of philosophy concerned with the nature of knowledge. Chapter 2 was an exercise in epistemology. Every theory has an, often implied, epistemology of knowledge. Most researchers will not deal with an explicit epistemology unless they are working with philosophical or religious materials. However, authors commonly introduce implicit epistemological assumptions into their arguments. For example, by assuming that all knowledge is from experience or conversely, that there are some forms of knowledge beyond experience (technically: a priori knowledge). This sort of division often comes into arguments over "values" or "social standards" and often ties to religious of ethical belief. (Religious beliefs contain elements that are beyond mere "experience.")

In practice, the assertion that some knowledge is "beyond" experience means that part of the theory are not open to empirical or rational criticism. It's important for researchers to note when this happens and what the author is doing with it. It can mean that there is something they do not want investigated or discussed. It can also mean that there is some element that their argument that requires you to "just accept" in order for their argument to work.

Assumptions

Theories also include assumptions. Assumptions are things that a theory accepts without proof. For example, some economic theories "just assume" that gain is the only motivation in economics. Assumptions can shape the theories and ideas based on them in very basic ways. For example, our current disciplinary divisions are largely based on 19th century assumptions. (This is one reason that the current divisions are getting so fuzzy!) Anthropology is "different" from sociology because non-Europeans are "essentially different" from Europeans hence two different disciplines. Political economy is separated into economics and political science because

practitioners believe economics "should" be separated
from politics, a basic assumption of "classical, free
market" economics. These particular assumptions have
had a major impact on how these disciplines developed.
For example, Adam Smith's "invisible hands" of the
"market" are often attached to legislators whose laws
structure the market. They are only "invisible" because
politics is formally excluded from the analysis of how
markets work.

Assumptions can also be very specific to the argument or
research. The immigration polices of Canada can be
"assumed" to be the same as the US (they aren't).
Welfare can be "assumed" to be a sign of social
dysfunction. (Is society taking care of someone old and
crippled dysfunctional?) We can "assume" that guns
"cause" crime and so on. In practice, all arguments and
theories carry a host of assumptions. In part, this is
because examining every "assumption" every time would
require immense amounts of time and effort. In part,
this is because authors usually do not make their
assumptions explicit, which makes them an excellent
place to hide or overlook things. (Identifying and sorting
out assumptions is a major part of a researcher's or
analyst's task.) In practice, we can not avoid assumptions
in our own or other's arguments. Nevertheless, we
should be aware that they are there. The question "what
is this argument or viewpoint assuming?" is a question
the researchers need to ask and ask often.

It's common for assumptions to be over-simplified or
partially true. Some authors use partially true
assumptions (like the welfare example) to gloss over the
fact that they are not making necessary distinctions.
Other assumptions are simply not true but are necessary
for the argument - which is why authors leave them
unexamined. For example, "guns cause crime" is

crediting a physical object as the "cause" of human behavior. Unless we believe that guns have mystical powers to influence people, this is simply untrue. Guns are involved with crime and with crime prevention, a much more complex relationship. However, simple "get rid of the guns" approaches to criminology require an equally simple causation to justify themselves, hence the need for the assumption.

Causation

This brings us to another part of theory, causation. This is sometimes called the "motor" of a theory (the part that makes it go). Scientists sometimes refer to causation as the "mechanism." Causation is extremely important because it tells us "how" and because it connects the various pieces of a theory together. It's often addressed with question like: "how does this work?" or "what makes this happen?" or "how are these parts connected?" It's extremely important for the researcher to sort out an author's chain of causation. (This causes that – which causes the next thing – which causes the other) One problem to be aware of is "gaps" in the chain of causation. Some authors will start with some very well established causation to establish credibility then make an unexplained jump (sometimes referred to as "then a miracle happens") to get to the conclusion they want. The researcher must examine each step to determine if the theory is actually in one piece.

Since research is often problem based, the identification of causation usually identifies were to address the problem. This makes identifying causation a central element in policy formation. The researcher must carefully trace through the chain of the argument's causation. Agenda scholarship and dis-information artists often conceal their distortions in the chain of causation. One needs to ask, "How exactly does this causation work?" If the researcher cannot answer this question then there's a serious problem with the theory.

We should also note that correlation is not causation. Just because two things are associated, say Scotland and red hair, does not mean that one causes the other. It's also very important to identify the direction of causation. This can get complex. For example, poverty and poor education are very closely related. Does poor education "cause" poverty or does "poverty" result in (cause) poor education. Alternatively, is this a cycle where one affects and "causes" the other? Which is primary and which secondary? Does POVERTY drive poor education or does POOR EDUCATION drive poverty? Does it vary with different groups – say by race or age? This sort of questioning is central to research. (Good researchers are people who can ask good questions. Indeed creating good questions is something of an art form and essential to analysis.)

In practice, solving this sort of maze is done by sorting out the different "causation chains" and subjecting each to empirical evidence. Then one has to integrate the various chains to see how they interact and try to reach a balanced overall conclusion. (You can see how my build and collapse theory would apply. As one worked through a dozen or more treads in this tangle each would yield new insight and data. Hence build and rebuild as one learns.)

Evidence

Theories are ideas - usually a bundle of ideas. In order to ground the intellectual world of theories in reality, the researcher must test theories against reality, against the "evidence." This is the process of relationism described in chapter 2. To do this a researcher looks for empirical evidence relating to their theory. (We can note here, that relationism works two ways, as a relationship between ideas (the theory) and how research is approached and as a relationship between ideas (the theory) and the empirical world.) These two relationships define much of research, which is both an intellectual exercise of the imagination and an investigative exercise in the "real world."

Researchers need to draw distinctions between different types of evidence. Evidence can be primary or secondary. Evidence can be hard or soft. Primary evidence is directly connected evidence. If one were researching the meaning of the Constitution, the Constitution itself would be primary evidence. Secondary evidence is something related but not directly connected. A comment by Abraham Lincoln on the Constitution would be secondary evidence. This seems simple but determining what is primary and secondary is not always easy. For example, a statement by Thomas Jefferson might be considered primary or secondary. Primary because he is the author or secondary because it's only his comments on not the actual thing measured directly. Since primary evidence is considered "stronger" than secondary evidence, it's important to be clear about what your evidence is.

Evidence can also be "hard" or "soft." Hard evidence is normally physical facts. Soft evidence is opinions or arguments. For example, the number of people on welfare this year and last year indicates if the number of welfare recipients are increasing or decreasing. This is hard evidence of a change in welfare rates. Commentaries by economists regarding how good or bad this is are soft evidence. The empirical measurements the economists used to support their opinions are hard evidence supporting their soft evidence opinion.

This is not as simple as it seems. In the natural sciences, the objects of study do not have a will of their own. In the social sciences and politics they do. This means that what you are measuring can be intentionally influenced. For example, numbers on welfare can be going down because regulators have tightened requirements. If you are using numbers on welfare as a measure of poverty, a simplistic interpretation of your numbers will make you believe the reverse of what is happening. (Numbers are going down due to regulations; poverty is going up as people loose their welfare payments.) The influence of people and policy on social "facts" can be extreme. For example, bank panics and runs on the bank used to be very common, with the introduction of deposit insurance they have become very rare.

Evidence can also be "adjusted' by careful selection of what is measured. Supporters of the Brady Bill (requiring background checks on gun purchases) claim success based on the 500,000 purchase refusals created by the law. Proponents interpret this to mean that the law has prevented 500,000 criminals from obtaining guns. This seems like "hard evidence" (an empirical fact) supporting the success of the law, but it rests on the assumption that all those refused were criminals. However, the "fact" of a refusal can also be generated by an administrative error or by the computer system being "down." In practice, the system itself is generating "facts" which may of may not have much connection to the original issue, in this case criminals buying guns.

The empirical evidence of refusals, which is taken to mean that criminals cannot get guns, also rests on the assumption that criminals cannot get a gun elsewhere.

Since this is demonstrably false, these very impressive numbers might not indicate anything other than the system's ability to make refusals. This sort of "evidence" can be identified by its tendency to clash with other hard evidence. In this case, the related statistic of seven actual prosecutions (a felon attempting to buy a gun is committing a crime), which gives a better indication of how effective the law really is. Problems with a theory's "evidence" can also be spotted because there are problem with its assumptions or with its causation. In this case, both such problems are evident.

(Author's note: you can see why I have so laboriously worked through ideas and principles. No amount of rules of thumb or "cookbook recipes" of analysis will work through this sort of tangle. One has to know how the system works and why it works that way. If you have a clear picture of that in your head sorting things out will come with practice.)

Another example of evidence manipulation is how categories are used. Money given to the poor is called welfare. Money given to corporations can be in the form of tax expenditures (tax expenditures are when the government doesn't collect part of the taxes owed. There are various methods, generically referred to as loopholes). In both cases, public money ends up in private hands. However, if you put these distributions of pubic moneys into different categories you can treat them differently. One can complain about the "burdensome and debilitating" payments supporting the poor, while also complaining that government is "not supportive enough" in payments to corporations. Direct comparison of these similar activities is prevented because they are in different categories. The handouts that "breed dependency" in the poor can "stimulate enterprise" in corporations and one is never challenged to explain why this is so different.

Another example of category manipulation is welfare fraud. This used to be defined as "intentional criminal" fraud. Now it is defined as receiving money while in violation of regulations. This seems the same but there are many complex welfare requirements, some very open to interpretation. Any technical violation, intentional or otherwise becomes "fraud." Clearly, "fraud" rates will skyrocket under the new definition. (When comparing numbers over time, or across jurisdictions, the researcher must be very careful that the definition or measurement method has not changed.)

Another problem with categories is the problem of mis-naming. By putting something into the wrong category, one can conceal its true nature. For example, some economists have declared that all unemployment is "voluntary." By doing this they can dramatically underestimate the problems of the unemployed – after all it cannot be that bad if it's voluntary. This misnaming process is often driven by the author's desire that readers accept the author's categories and terms. Use of these will channel the reader into addressing the issue as its been structured by the author.

Measurement Systems

The system of measurement can also affect results. (In the strict sense, systems of measurement will always affect results as in "the question determines the answer.") For example, school testing is increasingly being done through statewide exams. However, the logistics of marking large numbers of tests quickly, uniformly and cheaply pushes the system towards the use of machine markable multiple choice questions. Unfortunately, multiple choice questions are very poor at testing comprehension or critical thinking and very good at testing factoids. The result of implementing such "standards" is to push the educational system towards the teaching of factoids. This is a return to the days of rote learning and demonstrably the least useful form of "education." Over time, the measuring system and its standards will transform the educational system. (In science, this is called the Heisenberg effect.) Putting stiff penalties on teachers and schools that "don't perform," only increases the trend towards a rote learning training system.

Measurement systems can also set you up for "system's players." Before WWII the Royal Air Force used to measure fighter squadron "efficiency" by how many planes they had available to fly. This seems very logical and straightforward; the higher the percentage of planes "flyable" the "better" the squadron. However, high performance aircraft require several hours of maintenance for every hour of flying. Peacetime squadrons are also training formations and their pilot's abilities will increase – and decrease - with how often they fly. This sets the up the potential for a system's player. A squadron commander who wanted very high marks (and promotion) could simply stop all flying. The mechanics would catch up on maintenance, all planes would be "available" and his mark would be 100%. Unfortunately, the critical skills of his pilots would be deteriorating, seriously reducing the effectiveness of the squadron. The measurement numbers would indicate performance was going up, in the real world it would be going down. The researcher must be particularly careful with the problem of system's players when examining institutions and organizations.

The researcher must also consider the logic of the argument (or theory). Formally, logic is the study of the structure and principles of reasoning. This is the study of how one thing may follow from another. (We can note here that "logic" is a science in itself. What I'm discussing are some problems relevant to a researcher working with non-technical writings.) The two principle kinds of logic are deductive and inductive. Both are important for the researcher. Deductive logic starts with a statement (or theory) and then examines evidence to see if the statement is true in the particulars under examination. Inductive logic looks at the particulars and derives a general statement (or theory) from them.

Logic tends to build in chains of reasoning. A chain of reasoning is very similar to a chain of causation. (If Fido is a dog and you own Fido then Fido is your dog.) In practice logic, - chains of reasoning - are subject to problems of causation and assumption. A line of reasoning can be perfectly "logical" but untrue because it is based on a false assumption. In practice, serious flaws in logic tend to be obvious if clearly stated. (A good reason to suspect something that's not clearly stated!)

Authors who wish to hide problems with their argument's logic will sometimes attempt to conceal the problem by locating it in an unexamined assumption. Problems in logic also turn up as "gaps" between statements ("and then a miracle happens"). This "gap" is hurriedly skipped over in the hope the reader will not notice. Another method of filling a gap in a chain of logic (a chain of reasonable assumptions) is to fill the gap with something that sounds good but may not exist. This is very similar to filling in "gaps" in a chain of causation. (Outside of formal exercises in logic, it's hard to distinguish between chains of causation and chains of logic. For many non-technical writings, the chain of causation is the "logic" that holds the argument together.)

Statistics

Statistics are a branch of mathematics and can be very complex. However, in most cases the researcher will not be dealing with complex statistics. (I'm assuming you are working with general non-technical literature.) The first thing to know about statistics is that they are abstractions. Statistics are not "reality" they are mathematical representations of reality. For example, American families have an average of 2.4 children. Has anyone ever seen a .4 child? Statistics also deal with aggregates. This is very useful for looking at events involving large numbers of people but can be misleading. For example, a man with one foot in a bucket of boiling water and one foot on a block of ice has an average foot temperature well inside the comfort range.

Many authors use the anecdote or story to illustrate what they want to prove. Without statistics, it's impossible to know if this is a rare or common event. For example, the Columbine School shootings started a wave of "zero tolerance" crackdowns and increased police presence in schools. This was done as a "necessity" because of "unsafe" schools. In fact – statistically - the average American child was already safer in school than anywhere else including the home (about a 1 in 2 million chance of dying in school). Of course, we can also note the effects of aggregating here. There are some very unsafe inner city schools but they are relatively rare when compared to all schools.

This connection between anecdote (a particular case) and statistics also works in the other direction. We could have a statistic that the birth of "aids babies" in our city of 500,000 has increased by 100% in the last year. This sounds alarming but if the numbers have only moved from one to two cases, the increase is not serious. (Technically when the numbers of incidents is too low to be separated from random "background noise" the numbers are said to be statistically insignificant. In this example such low numbers could be just random chance and tell us nothing about the problem other than it exists.) To get a good idea of what is happening you need both the description of the anecdote and the statistics to tell you how common it is (technically: how often is "frequency", where is "distribution").

You will also note I was careful to state how big the city was (technically: the "population" of the study). This is important; the US has some 270 million people, in a "population" that large, even rare events can generate "large" numbers of cases. A one in ten thousand chance would give 27,000 cases; a one in a million chance generates 270 cases. Hundreds of cases may not mean that a national effort is required. This is especially true if a problem is concentrated in a particular area. For example, violent youth gangs are very tightly concentrated in inner cities. An anti-gang program for all schools would influence the vast majority of youth not in gangs; yet be too dispersed to have much effect on the actual problem area. Good statistical work tends to sort this sort of problem out.

In dealing with theory and its components, the researcher is dealing with both ideas and reality. On the one hand, the ideas structure how reality is perceived and measured. On the other hand, reality pushes back against our imaginings. The process of research becomes a dialectic (a "conversation") between ideas and reality. Intellectual structures are built; then empirical evidence is sought to confirm or deny these structures. This tends to be a back and forth cycle as one modifies the other. Eventually the researcher reaches an understanding of what is going on and analysis is achieved.

Dis-information

Dis-information is a special problem that muddies the water for a researcher. Since it's becoming so common, I'm going describe some of its common characteristics. Dis-information is a much stronger form of intellectual abuse then agenda scholarship. Dis-information includes deliberate attempts to prevent knowledge formation as well as promoting information intended to deceive. It can be a simple lie endlessly repeated until everyone "knows it" (welfare queens in Cadillacs) or it can be the manipulation of information to effectively create a lie. In dis-information, this distortion is intentional and suppression of the truth is the intent.

Dis-information campaigns are easiest when there is a large knowledge gap between the originator of the information and the recipient. The more authoritative the source and the more unsophisticated the receiver, the easier it is to make dis-information work. If the sender can also censor other sources of information, this will increase their likelihood of success. This is especially true if the dis-informer can hide the act of censorship. If people believe that their information is not censored they will believe that they are getting the full story and are much less likely to question the information they receive. If the receiver has no alternative sources of information and no way of forcing more information from the dis-informer then their ability to question the information is severely impaired. These three conditions of ignorance, coercion and impotence support the influence of dis-information. The Internet, by offering alternate sources and increasing the sophistication of the reader is a threat to dis-information campaigns.

Dis-information is not just about information denial like censorship. It seeks to hide the truth by side tracking investigations with false theories, planted "evidence" and efforts to break the construction of a chain of evidence that would reveal the truth. In other words dis-information seeks to subvert and side track the construction of knowledge and does so at every stage of the process. This is a much wider range of activity than mere censorship or lying.

There are a number of effective ways to hamper researcher's efforts. One of the most effective is to convince the researcher that they "don't know enough" or are "not smart enough" to understand what is going on. This can be done by producing a maze of complex data designed to confuse. It can also be done by convincing the researcher that any error in their progress "proves" that they have failed. (The construction and collapse of theories and chains of evidence is normal in research not failure. If you got it "right" first time every time you would be quoting dogma not researching!)

The first line of defense for a dis-information artist is to prevent the issue from ever becoming public. Simply being ignored obscures the matter. If critics raise the issue, you simply ignore them to preventing any public debate. This is particularly effective when they have a much greater access to the press then their critics. By refusing to debate or argue, the dis-informer denies critics the public exposure they need but can not get on their own. The matter becomes a non-issue and eventually is lost in the clutter of newer events. The Internet is a danger because it keeps issues alive. Information and new information keeps piling up in issue oriented sites. Online debates also keep the issue alive and can spark debates that are more public.

Dis-information can take the form of attacking critics. This can be done in several ways. The simplest is indignant outrage to indicate how ridiculous the criticisms are and how "inappropriate" it is to criticize the people concerned. This side tracks the debate into the "appropriateness" of criticizing this group and whether the critic is "morally qualified" to make such a criticism. This moves easily into criticizing the critic's motives. This again shifts the issue to the critics themselves and puts them on the defensive.

The person's motives can be conjectured, twisted or invented out of whole cloth. This is especially effective if it can be used to generate legal problems for the critic. Slander and extreme name-calling can easily be attached to an attack on motives. The purpose is to label (mis-naming) the critic as beyond the political pale. This both discredits the critic and discourages others from being associated with them. The classic example of this technique is "red baiting" were any critic is branded a traitor and subversive.

These methods are designed to distract the critic by emotionalizing the argument. Someone who has lost their temper is easily led off topic. An emotional exchange means that the actual issue is dropped and often forgotten. Critics can also be portrayed as "overly" concerned or as someone who has no case beyond their own emotional involvement. All of these methods deflect the debate from the real issues and so prevent the real issues from being addressed. These "attack the critic" techniques are particularly effective if you have more access to the mass media and the critic's reply is seldom heard. The Internet, by giving the critic an unlimited ability to present their own case in a web site gives them a voice. The potential for that web site to be visited by hundreds of thousands makes that voice dangerous. Critics hyper-linking to each other build on each other's work and create a new potential for organizing.

The dis-information artist also has a bag of tricks relating to their own behavior. One of the most prevalent tactics is to hit and run. This is a standby on the mainstream media. The argument is made or the viewpoint expressed, often as "just is" received wisdom and alternative views or criticisms are simply not heard or heard with far less "air time." Ample time is then available to ignore counter arguments or simply dismiss then without further debate. The owners of the mass media routinely use their ability to pick content and regulate discussion to shape perceptions. (If you think this is not the case surf across the channels during the evening news hour. You will find that supposedly independent, supposedly competing news hours all have the same stories all presented from very similar viewpoints. Out of the hundreds, if not thousands, of issues and news stories they seem to all pick the same dozen or so, all at the same time and lose interest all at the same time. The only exception to this is a smattering of local human-interest stories. In practice, this week in week out control of the news indicates an amazing amount of political conformity and intellectual consensus by supposedly independent news organizations.)

Another tactic of the dis-informer is to present themselves as authorities. This is particularly effective if

134

they fill their argument with details and secondary issues. This gives the impression that they have a host of "important things" at their fingertips and emphasizes their critic's ignorance. (Their actual level of ignorance, or not, is irrelevant if the dis-informer can prevent them from being heard.) A mass of irrelevant detail and side issues helps to muddy up analysis and opens up opportunities to change the subject. Currently, there is a growing tendency to use front organizations to establish "credibility." This can include bogus organizations whose real aim is to defeat what they are supposedly for by diverting resources from real organizations and discrediting their viewpoint by their behavior.

The dis-informer can also use their control of the debate to play dumb. No matter what evidence critics bring the dis-informer does not get the point or gets the wrong point – which allows them to change the issue. Distracting attention to another event also helps to silence critics, the more trivial or emotional the event the better. With media control, the dis-informer can decide what is "really important." This "really important" usually amounts to presenting their views and supporting their agendas. By threatening, this ability to ignore, sidetrack and substitute the Internet threatens those who wish to sidetrack debate and subvert political discourse.

A dis-information campaign can also be directed against the evidence that would allow a researcher to put together the truth. One way of doing this is to weaken the links in the chain of evidence or causation. If the links can be obscured, misnamed, dismissed or muddied up with false "evidence" analysis will be obstructed. The "vanishing" of evidence and witnesses is the ultimate form of "link destruction." There have been a number of national scandals where the "vanishing" of witnesses and evidence has been a separate scandal in itself.

Dis-information artists can obscure evidence by using a "straw man." A "straw man" is a position that is not actually held by one's opponent and distinguished by its extreme weakness. The critic is defeated by attacking this false position while claiming it's the critics. For example, one could argue that, a critic of "zero tolerance" in schools "don't care" about school safety. This tactic can be reinforced by insisting that the critic meet extremely high standards of proof. Any proof presented can then be challenged on its lack of "merit." This can be extended by "requiring" an absolute 100% solution to everything as a necessity to prove anything. Using these methods requires control of the media and debate. Without a fair degree of control the dis-informers ability to shape the debate in this way becomes problematic.

Sidetracking is another form of dis-information. This can be done with a well placed admission on some small or irrelevant point followed by the assertion that the matter is now "resolved." Attempts to continue to the real issues can be dismissed as emotional attacks, misguided or politicking. The matter can then be discarded as "old" and "dealt with." If the critic does not have the media access, to challenge this maneuver or keep the issue alive it then drops out of debate.

A variation of the "straw man" technique is simply to change the subject. The sensitive issues drop out and new "more important" and more manageable topics are discussed. This is particularly effective if the dis-informer has allies to pick up and continue the new discussion. When this is done formally, it involves a false investigation. This is controlled through control of the mandate, questions, staffing, or witnesses that are allowed to speak. The matter is then completely controlled, the "correct" findings can be made and the matter officially closed. If the records and evidence can then be legally sealed of destroyed so much the better.

Ultimately, this approach attempts to establish a "new truth" which then becomes "official truth" which critics must challenge. This can be done by inventing or suppressing facts and evidence, but is more powerful when it simply fits the facts to a new story. The matter is then "answered" and no further comment is necessary. This new story need not make perfect sense, for example, the "magic bullet" in the Kennedy assassination. Or, if no story can be invented, the matter can be relegated as an enigma too complicated and obscured to ever resolve.

All of these techniques of dis-information demand a certain amount of control over the debate, which in practice means control over the media of debate. The speed with which the government turned over the publicly owned and developed Internet to the commercial sector was a major political act and parallels the handing over of the public radio and TV media. This happened at precisely the point where the Internet was starting to become a mass medium and was done with little debate, comment or reporting.

For the first time in close to fifty years we are moving into an era where there is both a public mass media and no "war" to justify authoritarian control. The First World War established official government propaganda and the "Red Scare." The Second World War established complete media control and the Interventionist State. The Cold War is now over and "commie hordes" are hard to find. A new mass media is open to the public.

This means that new political, social and intellectual options are open to the country. But these options and their openness are not uncontested and it's not coincidental that the elites are feverishly pursuing new domestic "wars" on pornography, guns, drugs and crime. We have reached the post-Cold War era and intellectually it's turning into the era of InfoWar in Cyberspace.

Bibliography

Barnett, Correlli. The Collapse of British Power. Gloucester: Alan Sutton Publishing Limited, 1984.

Chomsky, Noam. Necessary Illusions: Thought Control in Democratic Societies. Concord: House of Anansi Press Ltd., 1991.

Collins, Randall and Michael Makowsky. The Discovery of Society. 4th ed., New York: McGraw-Hill, 1989.

Fisher, David Hackett. Historians' Fallacies: Towards a Logic of Historical Thought. New York: Harper & Row, 1970.

Lloyd, Christopher. The Structures of History. Oxford: Blackwell Publishers, 1993.

Mannheim, Karl. Ideologie and Utopia. Trans. Louis Wirth and Edward Shils, London: Routledge & Kegan Paul, 1936

McQuaig, Linda. The Cult of Impotence: Selling the Myth of Powerlessness in the Global Economy. Toronto: Viking, 1998.

Myrdal, Gunnar. Objectivity and Social Research. Middletown: Wesleyan University Press, 1969.

Reich, Charles A.. Opposing the System. New York: Crown Publishers, Inc., 1995.

Saul, John Ralston. The Doubter's Companion: A Dictionary of Aggressive Common Sense. Toronto: Viking, 1994.

Shah, Indries. A Perfumed Scorpion. London: Octagon Press, 1978.

Stauber, John C., Sheldon Rampton. Toxic Sludge is Good for You: Lies, Damn Lies and the Public Relations Industry. Monroe: Common Courage Press, 1995.

Strassmann, Paul A.. The Politics of Information Management: Policy Guidelines. New Canaan: The Information Economics Press, 1994.

The Squandered Computer: Evaluating the Business Alignment of Information Technologies. New Canaan: The Information Economics Press, 1997.

Tarnas, Richard. The Passion of the Western Mind: Understanding the Ideas that Have Shaped our World View. New York: Ballantine Books, 1991.

Wolf, Eric R.. Europe and the People Without History. Berkeley: University of California Press, 1982.

www.ingramcontent.com/pod-product-compliance
Lightning Source LLC
LaVergne TN
LVHW022323060326
832902LV00020B/3628